I0081626

GOLD CAMPS & SILVER CITIES

GOLD CAMPS & SILVER CITIES

Nineteenth Century Mining in Central and Southern Idaho

Merle W. Wells

2nd Edition, 1983

Published in cooperation with the Idaho State Historical Society and
assisted by a National Park Service Historic Preservation Planning Grant

Idaho Department of Lands
Bureau of Mines and Geology
Moscow, Idaho 83843

Bulletin 22

05-01-9518

CONTENTS

(Numbers in parentheses refer to index map on page 2)

PREFACE

More than a century has gone by since mining started in Idaho. Our perception of actual early day operations has been obscured by the long passage of time. Much has been written about mining in the West, but many Idaho camps have been neglected. Anyone wishing to know something of their development is pretty much at a loss to find such information. The mining history of only one of them has been thoroughly studied. That one is South Boise, which Robert L. Romig investigated during his graduate work at the University of California. His findings form the basis of the South Boise section of this account and have contributed in important ways to other sections. When more of these mining regions have been studied, a reliable impression of mining in the territory of Idaho will be possible at last.

In the meantime, this account presents some of the highlights of the development of southern Idaho's nineteenth century camps, as seen by the miners of the time. The miners' view of their camps differs almost entirely from the popular impression of today. Where tourists now see ghost towns, miners of a century ago saw stable, permanent communities, most of which they expected to last indefinitely. Where television watchers of today see vigilantes and gunfighters, old-time miners saw hard work coupled with the excitement of gold rushes and, if one of them was lucky, rich streaks of mineral in his own claims. The miners' account, then, is mostly one of hard work, enlivened by the excitement that went into developing those early Idaho districts.

This account is not a general history of Western mining or of life and society in the mines. Those stories are already available elsewhere. Rather, this survey covers each central and southern Idaho mining area to 1900. By that time, mining in Idaho had gone through several stages. An initial phase, in which the better placers were skimmed off, also had limited lode operations. After a halting, mostly uncertain start, Idaho's early lode miners went through an era of frustration and delay. After 1869, Idaho's placers provided work mostly for industrious Chinese miners. By 1880, lead-silver mining gained importance. Several new mining areas suddenly were developed. Finally, some scattered districts—rather modest in their early development—were located after 1884. By that time, rail service enabled lode miners to produce on a large scale in camps which had been retarded in earlier years. Each of these four phases of Idaho's nineteenth century mineral development had representative mining camps. They are presented in sequence of discovery or development, with major attention given to Idaho's gold rush decade prior to 1869. Summaries of later development are included, followed by a detailed account of Idaho's final major gold rush to Thunder Mountain. Because of extensive press coverage, and because Thunder Mountain provided an exceptional test of nineteenth century prospecting methods and development procedures, a large number of contemporary accounts of that challenge are presented.

Mining in Idaho brought permanent communities to a frontier area substantially sooner than would have been possible otherwise. These settlements had a remarkably diverse history, as is indicated in William S. Greever's *The Bonanza West: The Story of the Western Mining Rushes, 1848-1900*, yet they exhibited general western characteristics explained in detail in Rodman W. Paul's *Mining Frontiers of the Far West, 1848-1880*. Although some of Idaho's metal areas remained unidentified after 1900, most had been discovered by that time. Prospectors looked everywhere for gold and silver, and after a forty-year search, they had not missed very much. Their story provides an instructive chapter in Idaho's history.

ACKNOWLEDGMENTS

Idaho State Historical Society and Preservation Office staff members who have contributed to this bulletin include Robert L. Romig, whose long term investigation of gold and silver districts has provided an interpretive context as well as factual information essential for this account. Larry R. Jones, state historic preservation office historian whose research covers all of Idaho's mining areas, has compiled a mass of important additional material for enlargement of this bulletin to its present scope. Dorothy Moller, Doris Camp, Marj Williams, Karin Ford, Elizabeth Jacox, James Davis, Don Watts, and Judy Austin are responsible for a variety of valuable features of this publication, and Ernest Oberbillig, as State Historical Society consultant in mining engineering, cleared up a number of important matters necessary for a more adequate understanding of Idaho's mining history. Since 1970, a National Park Service historic preservation survey and planning grant program has funded major segments of this project.

Idaho Bureau of Mines and Geology staff members who have contributed greatly to this bulletin include Roger Stewart and Earl Bennett, in addition to Earl Cook who suggested and arranged for its original publication in 1964.

Photographs and illustrations were supplied courtesy of the Idaho State Historical Society.

Part I. First Gold and Silver Excitement, 1862-1869

INTRODUCTION

Gold and silver mining in southern Idaho from 1862 through 1869 went on in several mining districts prominent at the time, as well as in a number of others whose importance lay in the future. Those eight seasons brought the excitement of a major western gold rush and its decline. But there was more than one rush.

Gold and silver discoveries, some big and some small, lured miners to at least one district in each of those years. Yet the succession of mining booms was by no means limited to Idaho. All over the West, the same pattern prevailed: prospecting led to discovery and stampede, and finally to the formation of a new mining district, if there really was anything in the area to develop. This kind of pattern, varying for placer and lode discoveries, may be seen in the story of the dozen or more early mining areas of southern Idaho, which gained recognition either then or later.

Idaho's mountain ranges presented a genuine challenge to gold prospectors. Many went through mining areas without noticing important mineral resources. W.W. DeLacy took a major expedition up the Snake River, which he ascended into the area later to be Yellowstone Park without finding enough gold to interest anyone. John Stanley's company had better luck in Stanley Basin and Atlanta after a futile investigation of Bear Valley and the rough Salmon River mountain country. Stanley's success set off a wild, but unproductive, stampede to the Middle Fork of the Boise River, as well as an unsuccessful early rush to Wood River. A later fraudulent rush to Yellow Jacket showed that, as late as 1869, miners had penetrated difficult terrain, in which mining would later develop, but without discovering minerals of significant value. Thousands of other prospectors investigated ridges and streams barren of precious metals.

A few prospectors had sufficient talent, energy, and fortune to find new mining regions. They needed to know exactly where to search and how to test potential placer deposits or lode outcrops in order to find gold or other valuable minerals. As a result of their ceaseless efforts, some forty-four southern Idaho mining areas came into production by 1900, and a half-dozen more followed soon after. Mining in scattered districts — some of great importance and others of modest scale — brought settlement to Idaho and led to the state's admission to the Union in 1890. Many highly successful operations, as well as many failures, accounted for Idaho's early economic development. The series of early gold rushes between 1862 and 1869 provided a substantial foundation for the settlement of Idaho.

BOISE BASIN

Gold in Boise Basin is reputed to have been known to a trapper of the Hudson's Bay Company as early as 1844. But the discovery that set off Idaho's major gold rush did not come until August 2, 1862, when miners from Florence and Auburn, an Oregon mining town near Baker, found a district that eclipsed anything that had been discovered in the Pacific Northwest. Although little was done to establish some excellent possibilities for Boise Basin until late in 1862, the next season brought the greatest influx in the Northwest to these new mines, and the excitement continued unabated to 1864. But by 1865 and 1866, the Blackfoot commotion in Montana had captured the imagination of those searching for new mines. By 1868 any remaining surplus population in the Boise Basin was swept off to White Pine in northeastern Nevada. Boise Basin continued to produce substantially even after attention was diverted largely to other districts. Unlike some placer areas such as Florence, which were largely worked out not too long after their discovery, Boise Basin retained its prominence after the initial rush had ended. But the major excitement flourished primarily between 1862 and 1864.

More than twelve years had gone by after the California gold discovery in 1848 before mining began in Idaho. Intensive efforts to find new gold and silver regions in the West during those dozen years resulted in interesting discoveries in Colorado (then part of Kansas) and British Columbia in 1858 and in Nevada (then part of Utah) in 1859. Less important finds had also been made in Oregon and Washington. At last in 1860, E.D. Pierce, a California miner and trader with the Nez Perce Indians, managed to prove that gold could be found in the Nez Perce country. A rush to Pierce's new mines, quickly followed by the much bigger Salmon River excitement at Florence, led to a great expansion of the northern mines. Gold in Montana (then part of Dakota) and southwestern Idaho (then part of Washington) came to notice in the summer of 1862. Of the 1862 discoveries, Boise Basin turned out to be the most important.

Placer and quartz discoveries in Boise Basin had become almost inevitable by 1862. More than one exploring party came this way. On May 27, William H. Oliver reported from Victoria that his twelve-man party bound for Florence had found Boise placers that would not pay to work. The chances were excellent that some other Florence-bound group might have come up with something better in the Boise region, just as prospectors trying to reach Florence from the upper Missouri River were to locate paying mines that set off the initial Montana gold rush the same summer. Still another Boise discovery possibility arose from efforts to find the legendary Blue Bucket mines supposed to have existed in central Oregon. The alleged Blue Bucket gold find in 1845 — attributed to a misguided immigrant party temporarily lost between Fort

Boise and The Dalles — was strictly mythical. But by the summer of 1862 an ever widening search for the elusive Blue Bucket site was bringing goldseekers into present southwestern Idaho. Furthermore, the rush to Florence had enticed thousands of prospectors to Salmon River, and most of them had thereafter to seek their fortunes elsewhere. Radiating out from Florence, some of those surplus miners from Salmon River found good placers at Warren early in the summer of 1862, and within a short time others got as far as the Boise region.

Moses Splawn, an Elk City and Florence miner, credited a Bannock Indian with the discovery of gold in Boise Basin. While at Elk City and at Florence, Splawn got acquainted with an Indian who, he reported, showed considerable interest in the metal that the White Man had come so far to

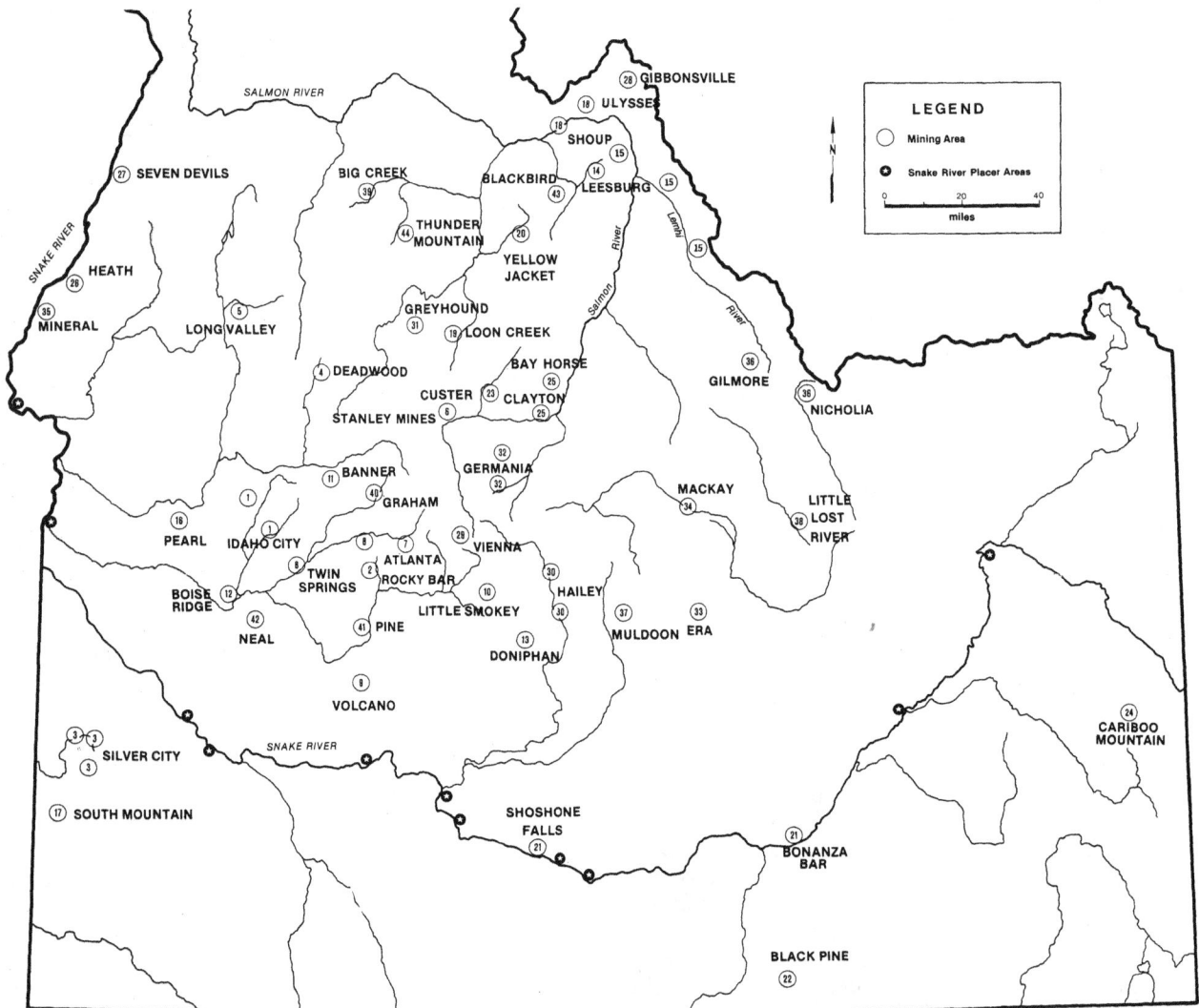

Index map of southern and central Idaho mining regions. Numbers in circles refer to mining areas listed in Contents.

find and had gone to such effort to get out from remote spots such as Florence. Later, Splawn met again with his Bannock friend along the Salmon River at Slate Creek while on his way out for the winter. At this point, the Indian suggested that if white men were so eager to find more gold, they ought to look in another basin farther south. Convinced that the Bannock had indeed seen gold in abundance in that southern basin, Splawn got careful directions and set out to raise a band of men to find the Bannock bonanza. Reaching Auburn (a new Oregon gold camp near present Baker), he joined one of the Blue Bucket parties that wanted to try the Owyhee country. When that hunt failed — although D.H. Fogus did find a little Owyhee gold on June 28, 1862 below Diamond Gulch on Sinker Creek, but not enough to warrant mining at that point — Splawn, Fogus, and a few others set out for the basin which the Bannock Indian had described as being across the Snake River to the north. On their way, they met a party under George Grimes and talked them into attempting to get across the Snake to try Splawn's venture.

Crossing the Snake River, even at the regular Oregon Trail ford at Fort Boise, posed a major difficulty that summer. An extremely hard winter throughout the Northwest had been followed by the record flood of 1862 down the Snake River, a record which has not been approached since. About a month was required to prepare for the crossing. Finally, the resolute prospectors were able to come up the Boise River to the canyon, where they turned northward along Boise Ridge. Although they had to go to some effort to avoid hostile encounters with numerous Boise Indians, who were expected to resent this intrusion into their country, the Splawn-Grimes-Fogus party finally reached its objective. At Boston Bar near Centerville on August 2, D.H. Fogus began to prospect again, this time more successfully. And Moses Splawn was convinced that these were the mines that his Bannock Indian friend had told him about.

Indian hostility, or at least the threat of it, kept the Boise Basin discovery party nervous. But the men continued to examine Grimes Creek up to Grimes Pass, where Grimes himself was shot on August 9. Although a strong tradition persists in Boise Basin that the Indians had nothing to do with the shooting, those who returned to Walla Walla credited the incident to a disaffected Bannock or Shoshoni. In any event, Grimes was hastily buried in a prospect hole and his men hurried back to the Boise River, where they happened to meet Tim Goodale's party of Oregon Trail immigrants.

Goodale, a trapper who knew the country, had

Centerville

just led an immigrant band over a new route from the Blackfoot River via the Craters of the Moon and the Camas Prairie. The group had been hopeful that he might be able to take them directly to the Salmon River and Florence. The best he could do for them was to open a route past later Emmett and Cambridge to Brownlee, where a ferry soon was established. Goodale's Craters-Camas Prairie route had been popular with fur traders many years before, but it had rarely been used by immigrants on the Oregon Trail or by prospectors. Protected by Timothy Goodale's large group, the discoverers of the Boise Basin mines had no more Indian trouble.

⚒   🐴   ⚒

A major gold rush to Boise developed in the fall of 1862 when further prospecting began to suggest the extent of the new placer country. The return party reached the basin on October 7, 1862, and established Pioneer City (usually known as Hogem in the early years) and Idaho City (known as Bannock City, or West Bannock, until the Idaho legislature changed the name to avoid confusion with another Bannock City in the territory that had grown up around claims discovered only a week before the Boise Basin find).

The first week's prospecting yielded about $2,000. In a letter dated October 14 from Grimes Creek to J.V. Mossman, William Purvine rated the new placers at about $18 a day. Indicating that they were more extensive, if not as rich as the ones at Florence — "eighty dollars to the pan being the greatest amount taken out at any one time," he noted also the "good extent of mining ground and abundance of timber, water and grass, in and about the diggings." Lyman Shaffer, a former member of the Washington legislature, wrote at about the same time that "some of the boys have got to rocking in choice spots on the hillsides and make from $8 to $16 per day to the hand, and sometimes as high as $20, carrying the dirt fifty and one hundred yards."

Little was known yet of the extent of the placers, but by the beginning of November hundreds of miners were leaving the Powder River mines for the new basin. Lewiston's *Golden Age* on November 6 reported that over a hundred already had left Lewiston. J. Marion More, elected from Shoshone County to the Washington legislature, arrived in Walla Walla on December 5 on his way from the new mines to Olympia, the capital of Washington, and admonished those who would join the Boise rush to wait until spring. Yet More reported that a claim near Placerville had been returning up to $300 per day and that Lyman Shaffer had found a good quartz lode in the vicinity. Many small gulches and streams paid from $5 to $50 a day. The future certainly was bright. Claims were yielding $40 a day to men who were willing to pack gravel on their backs to water, and some were good for $100 to $200 per day. In December 1862, mining was carried on at such a disadvantage that only a few exceptionally spectacular claims were even worked at all.

Disregarding the inconvenience of a winter trip from Lewiston or the Columbia River to Boise Basin, thousands of hopeful prospectors joined the rush to the Boise mines by the end of 1862. Anyone who might have delayed until spring could have anticipated almost with certainty that the men who had rushed into the new mines in the winter would get all the good claims, although these newcomers could not expect to do much more than get established and prepare for the spring season. Under the circumstances, hardly anyone could affort to wait. William Pollock's experience on Granite Creek (which he reported on January 9 to the *Golden Age*, dated February 15) is a sufficient illustration. His party of six—five of whom were experienced '49ers—heard of the new discoveries on the stream where they camped the night they got to the basin. The next morning, each took a 200-foot claim and began to prospect. Pollock reported:

> We immediately sunk a hole to the bed-rock; the prospects were not very flattering until we reached it, and then the first shovelfull made our eyes stick out, for the real stuff was there in all its beauty. We panned out three pan-fulls of dirt, and the proceeds were $11. That satisfied us, and so we filled the hole up with water and dirt, and tried to get more ground, but it was no go, for before we finished prospecting the men poured in so fast that the ground was all taken up; our eyes stuck out pretty big, but after all we were satisfied. We have six claims of 200 feet each, running up to within 70 yards of a rich quartz lead, so they say; at any rate it prospects well, and looks as though when tested it would prove almost solid gold. . . . You could not buy any one of our claims for $3,000 down, and as for mine, I would not sell it for any price, as I think all the money I want is in the claim. The gold is generally coarse, about the size of a kernel of corn, and from that to $4 and $5 pieces. I think there is about 2,000 men on Granite Creek or near here, and thousands more will find good diggings here in the Spring. The snow at present is about four feet deep, and keeps acoming. We cannot work much, but we find plenty of pine trees for fuel, and plenty to eat and drink.

Had there not been so many prospectors around, Pollock's crew might have taken up a number of additional claims. Altogether, local mining law in the basin allowed each man to hold a total of five claims—one creek, one wet gulch, one dry gulch, one hill or bar claim, and one quartz claim.

Population reports coming from the basin at the beginning of the year showed that by January 12 Placerville had a population of nine hundred to one thousand and seventy-five to one hundred houses already built. Figures which reached Lewiston in time for publication in the *Golden Age* on January 29 showed that four thousand men had already found their way into the basin, hundreds more were arriving daily, and seven to eight thousand were expected by mid-February.

Although very little mining actually was possible, excellent initial returns kept the gold seekers there in a frenzy of excitement. Some claim

J. Marion More

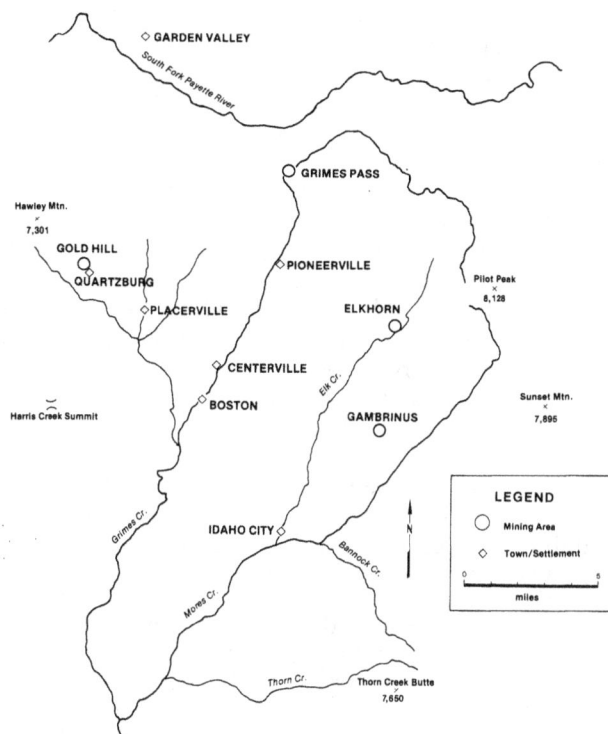

Boise Basin

owners on California Gulch were making 25 cents to $4.50 a pan; eager miners found another gulch to be "immensely rich — paying from $15 to $20 to the floursack of dirt, which they carry on their backs to the water." A man who wrote on February 15, 1863, from Idaho City to the *Oregon Democrat* (March 23, 1863) cleared $3,000 over expenses in the occasional times that he could mine during the winter. He expected to clear a total of $20,000 over his expenses before he got the claim worked out. And he could expect one or two years to go by until he would have a chance to get it worked out. Reports such as these kept the rush to Boise going strong. The *Golden Age* of February 25 published this letter from a Boise miner:

> I have been here some three or four days, and find things very brisk. These mines are going to be the richest that have been struck since '49; in fact many old Californians say they are a great deal better. They are in a basin some 30 miles long, by 25 miles wide and every gulch and creek pays well — the dry gulches pay from 50 cents to $5 to the pan. I have been prospecting the last two days and have got fair prospects. This is the richest country I ever saw. I would not take $10,000 for my prospects. I have taken 200 feet in a quartz claim, and have got a good creek and gulch claim all paying well. There are three mining districts in this basin; one is called Moore's Creek, the other Grimes; and the other

Placerville; they are very large camps. The miners are making from $50 to $60 per day, and packing their dirt in flour sacks from 100 to 200 yards.

Reports of astonishingly rich claims kept the rush to the Boise mines going full blast. By the middle of March, a Lewiston correspondent noted that "large numbers are leaving Lewiston daily, in small boats, for the Boise mines, that being the cheapest and easiest route. A great many are leaving by land, also. One hundred loaded pack animals left Lewiston on Monday last." He continued to say that although little mining was possible at Placerville, except with rockers, the results "give rise to the wildest speculations as to the extent and richness of the Boise mines. Men write about their claims yielding from fifty to one hundred dollars per day, and ounce diggings are scarcely mentioned."

Preparations made during the winter paid off during the short mining season in the spring of 1863. Ditches were dug and sluices installed for a large share of the hill and gulch claims around Placerville, in anticipation of the season that turned out to last only about ten weeks or so. Melting winter snows made early spring placering possible, and gulch claims averaged about an ounce a day during the better part of the season. But getting water was a problem before April was over, and by late May many of the best mines in the Placerville vicinity had suspended operation. A more extensive network of ditches had to be constructed to overcome the water supply difficulty.

Miners on Mores Creek around Idaho City were better off. With eighty miles of ditches ready for the initial season, and water expected to be available all summer, they did well indeed. Aside from sluicing, one enterprising company of miners had already installed a hydraulic giant, the only one in operation east of the Columbia River, on the north bank of Elk Creek. Five men using it were able to average 100 ounces a week. The richest claim around Idaho City that summer was the Idaho, owned by seven men and containing 2,000 square feet running right to the edge of town. Bedrock on it averaged a dollar a pan and ran as high as $9.25. The owners employed fifteen to twenty-five men to make ditches and operate the claim. Claims that paid less than $8 a day were not even worked, and many of the miners preferred to work for $6 a day to develop their own claims.

With the advent of summer when water gave out, activity did not cease in the basin, even for areas where gold production was halted. Towns

Hydraulic giant at Idaho City

had to be built and preparations had to be made for future mining seasons. Knowing that operations would go on for years, rather than come to an early sudden end (as had been the case in Florence where most of the best placers had been worked out in one grand season in 1862), miners, merchants, artisans, and professional people set out to establish a stable community, later to become the major gold center of Idaho. Indicative of this development, a new stampede from Lewiston — this time of suppliers with one thousand or so packhorses — hit the trail on June 16, 1863, to the Boise Basin mines with heavy loads of provisions, clothing, and tools.

By this time, there were plenty of miners to be supplied. Placerville and its suburbs had a population of two thousand, out of a total of twelve to fourteen thousand in the basin. Placerville had been the leading city of the basin with eighty-seven frame and log houses, thirteen saloons, five blacksmith shops, seven restaurants, and five meat markets which dispensed two tons of beef daily. Placerville's location on the west side of the basin made it the first stop for miners coming into the region and the original supply depot for the basin as well.

With the general end of the mining season around there, population shifted to Idaho City (then called Bannock City). By July 20, that newly expanding center, according to a report in the *Daily Alta California* (San Francisco), had "two main streets, each a half mile long. Hotels, restaurants, and grog-shops abound. All the town has been built since last winter. Crime is frequent, homicides occur most every day. There are no churches or

schools." Another correspondent, writing only two months later on September 25 to the *Sacramento Daily Union,* noted the rapid change in Idaho City which now had a printing office, eight bakeries, nine restaurants, twenty-five to thirty-five saloons, forty to fifty variety stores, fifteen to twenty doctors, twenty-five to thirty-five attorneys, seven blacksmith shops, four sawmills (two of which were steam), two dentists, three express offices, five auctioneers, three drugstores, four butcher shops, three billiard tables, two bowling alleys, three painters, ten shops, one photographer, three livery stables, four breweries, one harness shop, one mattress factory, two jewelers, and a dozen other assorted businesses. Building lots had considerable value, ranging from $500 to $2,000. Possibly part of the reason for their value was that they were located on excellent quartz ground. Conditions in Idaho City had improved substantially:

> On Sundays the miners swarm into town from their camps, and render the day exceedingly lively, especially as they purchase most of their supplies then, and do up their drinking for the week. Society is rapidly improving here, owing, doubtless to the influence of women, a large number of whom are here and perhaps to the promptness and vigilance of our officers, who leave no stone unturned to ferret out and bring to justice offenders against law and order.

By the middle of September, Idaho City with a population of 6,267 (of whom 360 were women and 224 children) had surpassed Portland to become the largest city in the Northwest. Placerville with 3,254, Fort Hogem or Pioneer City with 2,743, and Centerville with 2,638 were all far larger than any other cities in the area that is now Idaho. Furthermore, the basin mines had led to the establishment of Boise as a service community. Founded in July

Main Street, Idaho City

Idaho City, 1875

Pioneer City

that there are claims enough to furnish profitable employment for 15,000 persons but owing to the scarcity of water at present, now more than one-third of that number can get work. None but creek claims and a few bar claims can be worked with sluices, though many persons are still at work in the gulches, some with rockers, and others preparing dirt for spring freshets.

With placer possibilities limited by lack of water, mining excitement in September of 1863 turned abruptly to quartz. Since late in 1862, important quartz properties had been known in Placerville, and soon some very encouraging lodes were attracting interest near Idaho City. G.C. Robbins, a former mayor of Portland, left the basin in September for San Francisco, taking along four hundred pounds of ore from a forty-foot shaft. While sinking the shaft, he had panned $1,470 out of the eroded outcrop, and his prospect looked promising indeed.

Lode development, however, took a long time. While quartz promoters were out trying to raise capital and arrange to bring in machinery to work

near Fort Boise, which the U.S. Army had established the same month along the Oregon Trail, Boise had a population of practically a thousand itself. In comparison, Pierce had gone down to 508, Florence to 575, Warren to 660, Elk City to 472, and Lewiston to 414. With a population of sixteen thousand in the basin, about half of whom were probably miners and the other half merchants, artisans, and professional people, there could be no question as to what were regarded as the important mines that year.

Mining that continued through the summer and into the fall of 1863 became chiefly creek mining, although special problems handicapped those who turned from valuable, but seasonally unworkable, bars to the stream beds. A correspondent writing on August 30 to the *Eugene Review* explained why bar claims were regarded as the most valuable:

> The creek claims are equally rich, but it requires four to eight feet stripping to get the pay dirt; besides cutting longraces, rigging pumps, etc., to keep the water out, as they usually are on a level with the bed of the streams. The mines are much varied in richness, some claims paying $100 per day, to the hand, while the greater part pay from only $5 to $20. Larger sums have, in some instances, been taken out but they are only occasional; no claims within my knowledge, average above $100 to the man. It is estimated

Placerville after 1880

Placerville in 1884

their properties, a mild late season coupled with fall rain allowed placer operations to resume. A surprisingly large fall production continued to December, and the 1863 total yield for the basin may have reached as high as $4 to $6 million. Although many more men had worked in the Boise mines in 1863 than had worked in Florence in 1862, the short season probably reduced the total yield for the basin's first full year to something less than the corresponding recovery for Florence.

⚒   🐴   ⚒

Too pleasant a winter, especially for lack of heavy snow, made for a discouraging outlook at the beginning of 1864, and most mining had come to a halt by February. Occasional chances to work through the winter made it possible for some of the miners to make something more than expenses. The *Portland Union* of May 5, 1864, for example, referred to an informant who had managed to average about $100 monthly over his mining costs even though his working days were somewhat infrequent during the winter. But even though most mining was largely shut down, the rush kept up. A writer from Centerville on March 27 complained in a letter to the *Trinity Journal* of April 30, 1864, that five or six hundred hopeful prospectors arrived in the Boise mines each week even though only a few of them could expect to get work. Another pessimistic correspondent wrote on April 30 from Idaho City to the *Sacramento Daily Union* that although the Boise mines were good, a severe shortage of water was expected to bring the main season to an end in about ten weeks, and after that not more than four thousand miners would be able to work. Creek claims could accommodate the four thousand, but employment for only that many men would not leave nearly enough jobs to take care of the thousands who were coming in. Still another report — this one hopeful that there would be more water than there had been the season before — suggested that the spring runoff was washing out and ruining much of the advance work that had been done in preparation for creek mining.

Regardless of the water problem in the Boise mines, the rush gained momentum in April 1864. Information that a host of Boise-bound miners were pouring north through Yreka, California, was confirmed by Charles C. Dudley, a prominent Californian who settled in the basin early in April. An extremely dry winter was driving miners out of northern California by the thousands, and Boise was the place to go. Most of them arrived when the season was at its height; at least production figures for the richer claims were still high. On one of the

claims, six men realized $1,000 in a twenty-four-hour day; in another, six men made seventy-two ounces in a twenty-four-hour period; on a third seven men recovered thirty-five pounds of gold in nine days; and on yet a fourth, four men in six days made $1,075. The basin camps were roaring as never before. This time the water pattern was known for sure in advance, and during the season in which water was available, mining went on uninterrupted for twenty-four hours a day, day after day. The *Boise News* of April 30, 1864, described the placering at night in Idaho City as a Grand View.

> A night scene in the Boise mines is as brilliant and magnificent as any similar spectacle to be met on the green earth. We counted more than thirty mining fires on Tuesday evening from a single standpoint in front of our office door. The ringing of shovels as the auriferous gravel slides from the blade, is distinctly audible above the murmur of the water in the sluices, conspiring with the haze and smoke through which the mountains beyond are dimly visible to render the scene most interesting and lovely.

Idaho City still was growing, boasting two hundred merchants by June 10. Two handsome new theatres were finished, ready to provide dramatic entertainment for the miners just as soon as anyone had the leisure again to be entertained. Meanwhile, miners were "mining right in the city under the houses and in the street." Montgomery Street, in fact, was panning at $16 a pan. Some bitterly contested lawsuits arose over mining in the city. Those whose houses collapsed after being undermined by placering may have felt that they had good reason to resent the intrusion of miners right into the town. The *Boise News* of June 11,

Main and Commercial streets, Idaho City, April 1894

Two views of Idaho City before 1900

1864, described how business properties were flooded as a result:

> The tail-race in front of this office, not having sufficient fall, is prone to fill up with sand, causing the bank to overflow occasionally to the annoyance of the denizens of Wall street, who have in more instances than one, of late, found muddy water running about their doors and filling up their cellars. The flume across Gardner's gulch is particularly given to overflowing, and when it does so, the Clerk's and Probate Judge's office becomes almost a floating palace. The Celestial laundry had also been annoyed from the same cause; on one occasion we observed a brawny son of Boodh endeavoring to stay the flood and turn it back upon its fountain with a clapboard, but the incarnate Guadama in *nir vana* was either powerless, or the disciple's faith too weak to prevent the water according to the laws of nature, from finding its level.

Before the water supply began to fail—at Placerville early in May, when Wolf Creek was getting so muddy it would scarcely flow anymore, and at Idaho City by early June—mining was definitely profitable. For example, in one of the better mines near Idaho City in an average week (in this case, the one ending May 22), from the yield of $14,750.50 was subtracted $6,979.36 for expenses which left $7,771.14 as the net profit for the four partners involved. As late as the beginning of June, fully nine-tenths of those around Idaho City who had a head of water as much as half a sluice in depth were making an ounce or more per man each twenty-four-hour day—that is, $160 per sluice. For each sluice, water cost an average of $25 to $30 a day, and labor cost $60. That left an average profit of $75 to $100 to the claim owners. By July 9, however, the Gold Hill hydraulic giant above the mouth of Bear Run at Idaho City was the only one apparently still going. Production for the basin had not ended anywhere, but the time had come for many of the surplus miners to move on. At the end of the 1864 placer season, it was evident that the best claims were still less than half worked out.

⚒   🐎   ⚒

Lode mining, which was not dependent upon a large water supply, began to get started in 1864, particularly after the placer season had mostly closed. Some recovery was made with large hand mortars that summer on the Landon mine near Idaho City: ore out of a thirty-foot shaft produced a modest $10 to $12 daily—the result of pounding up one or two hundred pounds of rock each day—until a total of $3,500 to $4,000 was recorded. Further

work on the shaft was held up pending the arrival of a stamp mill. The plan was to work the Landon in conjunction with the nearby Gambrinus (which had been incorporated for $380,000 on April 14, 1864); both mines had been put under the joint management of Albert Heath. As was the case in financing practically all quartz mines of the time, Heath lacked the capital for development (and perhaps even the knowledge that development was necessary) prior to installing a mill, and he made little or no effort to find out in advance just how big a mine the Gambrinus-Landon property actually was. An arastra on the Gambrinus had been yielding $200 a ton that summer, and surface prospects certainly looked good.

When H.H. Raymond's ten-stamp Pioneer mill arrived at Granite Creek near Placerville on September 15, 1864, it looked as if the day of lode mining had finally dawned in the basin. Part of the Gambrinus mill arrived on October 16, and $263,000 came from that property in 1864 and 1865. Other mills were expected to be in operation by the next year. Crushing ore from the Gold Hill, Raymond's Pioneer mill turned out $5,000 in its first week of operation; succeeding mill runs were regarded as better still, for a short time at least. With ore from a 130-foot tunnel, Raymond's San Francisco Company produced $20,000 by the next spring when the mill suspended operation and work ceased. "By a course of speculation and mismanagement," Raymond was thought to have ruined the enterprise, and eventually his mill wound up becoming a sawmill.

Even without Raymond's failure, the spring of 1865 was something of a disappointment anyway. To begin with, the gold rush from Portland was deflected to Kootenai in British Columbia, where placers had been discovered on March 14, 1864, and then to Blackfoot in Montana. Furthermore, an extremely hard winter had set the whole country back. Then on May 18 a fire broke out in Idaho City, the first in a series of four such disasters, and destroyed over a million dollars in property. Although the best production had been realized by that time, the basin placers were by no means worked out. The town was rebuilt promptly, but a great oversupply of goods had been rushed in by merchants wishing to capitalize on the shortages created by the fire. When the customers failed to materialize, many had to sell at a loss. The miners may have benefited by the fire, however, both in lower prices and in a chance to placer out much more of the burned-over area of Idaho City.

Not only did placer mining continue to flourish in the basin in 1866 and 1867, but also quartz mining kept on after Raymond's initial failure. The

Gold Hill had placer as well as lode mining

celebrated Elk Horn mine (which among other things seems to have tempted the Boise County treasurer to invest the county funds in it, with rather unfortunate results) finally produced $9,105 in the fourteen days ending April 28, 1867, $8,200 more in a run ending June 15, and another $12,000 in a cleanup July 9, 1867. An eighteen-inch-wide, free-milling vein, developed with stopes above a fourteen hundred-foot tunnel, accounted for $500,000 worth of gold mostly by 1868, with intermittent work in later years.

In the spring of 1867, William M. Lent and James M. Classen reopened their Pioneer property on Gold Hill. Before installing their mill and beginning to mine, they were wise enough to have the ore tested and some of the mine's extent determined in advance, and to prepare adquate cost estimates of machinery and operations. Their example in undertaking at least a limited amount of development before going ahead with their mill was notable as one of the very few instances in which

quartz men in the basin operated by the proper, but often disregarded, mining methods known at that time. Lent and Classen engaged "the noted and experienced quartz miner," George Hearst. Hearst, then making a tour of the basin and collecting specimens from the leading lodes, was already known as one of the most prominent California lode miners. (Eventually his mining fortune, of which a modest part came from Idaho mines, laid the foundation for his election to the U.S. Senate and for the newspaper empire developed by his son, William Randolph Hearst.) Satisfied with preparing adequately for a quartz mill in the Gold Hill, Lent and Classen acquired for a modest price the twenty-five-stamp Chickhoming mill, which Richardson's New York company had brought to the basin at a $75,000 cost but had run only long enough to test it.

Powered by an eighty-horsepower boiler, the mill was expected to accomplish great things on the Gold Hill. Twenty-eight men were employed in running a new tunnel and operating the mill. Successful management of this enterprise was expected to restore confidence in Boise Basin quartz. But in spite of apparent confidence, the venture did not attain the success that was anticipated. The Gold Hill had to wait until the fall of 1869 for production to commence. The milling continued with only slight interruption for the rest of the century and beyond.

During the fall of 1867 Ben Willson, O.G. Waterman, and William Law extended a big ditch from their large operations at Grimes Pass toward Centerville. Running on a new and higher grade, this ditch made it possible to deliver water to many claims that had been in short supply before. It was large placer enterprises such as this one that helped to keep the basin going. But 1868 proved nevertheless to be a relatively dull season.

Because operations were large scale and efficient enough that comparatively few men with hydraulic giants could work a great amount of ground for high returns, the host of miners who once had swarmed over the basin were of necessity leaving for new discoveries like White Pine in northeastern Nevada in 1868 or Loon Creek in the upper Salmon River country in 1869. Chinese were coming into the basin in 1868 to fill up some of the gap. By the end of 1868 some of the important basin placers, particularly those around Pioneer City, were beginning to be worked out. James O'Meara noted: "It is useless to disguise the fact that Pioneer has seen its best times and that it is rapidly on the decline in trade and mining importance." Although it was still expected to be a fair camp, with some good ground remaining, there was no doubt that its production in the future no longer would compare

Quartzburg

Idaho statehood celebration at Quartzburg, July 4, 1890

with that of 1867 and 1868. Some good water seasons had greatly speeded up work.

The situation around Placerville was more encouraging though. Some of the best ground in the basin remained there, largely because it could be worked only a few weeks each season. Moreover, the means of working the Placerville mines, along with those at Centerville, were at hand. With a good water year in 1869, new large ditches supplied twenty hydraulic giants within a mile of Placerville.

Yet, even by the end of the rather successful 1869 season, many parts of the basin had been worked only partially, and reworking needed to be done upon many of those where an initial job had already been completed. As a consequence, Boise Basin continued to operate for many years as both a placer and a quartz camp.

Ambitious plans to supplement basin placers with lode production included installing in 1869 a smelter in Pioneer City to handle silver galena.

Quartzburg in 1899

Last Chance mine at Quartzburg, August 1900

Quartzburg (c. 1900)

More successfully, the Gold Hill mine managed to operate from September 1869 to July 1876 without significant interruption. In 1873 when it adopted the Washoe process, the $300,000 recovered so far represented only five percent of the mine's eventual ore value. The Gold Hill had saved its sulfide tailings for later reprocessing, and ran into no serious difficulties until the summer of 1876 when the operation got below water level. After a drainage shaft was dug, the Gold Hill complex continued year after year to grind out thirty-five to forty tons of ore daily. In 1886 the Gold Hill hoist was destroyed in the Quartzburg fire. Mining resumed after reconstruction, and by 1896 total production amounted to $2,225,000. Even in 1870, when placer production was declining abruptly with a preponderance of Chinese miners who worked low yielding claims, lode values did not account for too significant a part of mining activity. Compared with Boise Basin's $3 million yield mostly from placers in 1870, Gold Hill's yearly value of less than $100,000 seemed modest enough. Yet the Gold Hill mill started off far ahead of the nearby Mayflower, which depended upon an arastra for processing ore from 1868 to 1870.

Placer values for the Boise Basin declined after 1870, when a $3 million recovery (compared with

The Daly Hotel in Quartzburg (c. 1900)

$842,000 for Owyhee and $350,000 for Leesburg) accounted for most of Idaho's mineral production. A shortage of water in 1874 reduced the basin placers to less than $500,000, but that collapse was not typical. Occasional new lode claims, such as the Forest King which had a ten-stamp mill and nine hundred feet of tunnel in 1884, nine years after discovery, helped keep up mining enthusiasm. So did the use of hydraulic elevators which allowed Pioneerville miners to work adjacent flat stream bottoms that could not be sluiced. Years of effort also went into the promotion of a bedrock flume project in Boise Basin designed to uncover flat bedrock where placer values were concentrated in otherwise unworkable deposits.

Eventually twentieth century dredging enabled placer miners to get much of the gold that had eluded their nineteenth century predecessors. Dredging raised total production in the basin to about three million ounces of gold, worth more than $66 million at the time. With the rise from $20.67 to $35.00 an ounce in 1934, Boise Basin's gold value increased to more than $100 million by 1942, when mining ended because of the war. In 1980 when gold prices exceeded $600 an ounce, a refiguring of the value in Boise Basin's production rose abruptly to $1.8 billion. Idaho's old miners never suspected the extent to which their mineral wealth would appreciate.

Gold Hill miners in 1915 and after 1920

Gold Hill and Iowa mill near Quartzburg (1912)

Boston and Idaho dredge, Boise Basin (1913)

Fires continued to destroy mining towns as late as 1899 or 1900, when this relief expedition set out from Centerville to assist Placerville fire victims.

Ben Willson's hydraulic elevator at Pioneerville (c. 1884)

SOUTH BOISE

Anxious to ascertain just how large a region the Boise mines would cover, and always eager to find something even better than the rich ground already known in Boise Basin, an impatient group of hardy prospectors set out to explore the country farther up the river long before the higher ridges and streams were free from snow. Rumors of rich placer possibilities at South Boise reached the Powder River mines of Auburn, Oregon, by April 1863, and in less than two months, by May 7, the placers had been traced up Feather River to some still more promising quartz leads in Bear Creek. Word of the new South Boise lodes, backed up with some "very rich" specimens, set off a stampede of some fifteen hundred Boise Basin miners on May 20.

After a hard trip over rough country to the new Eldorado, most of the fifteen hundred rushed right back again. Although they had found high-yield placers on Red Warrior Creek, where more than one hundred claims were taken up in May, and some good ground on Bear Creek near the quartz outcrops, the new placers were not nearly extensive

enough to hold the horde that had joined the rush. Furthermore, the promising quartz prospects could not be developed for a season or two at best. The better Red Warrior claims were good for $20 to $25 a day in May. A few sluices actually in operation by the middle of June ranged from $16 to $60 per day per man. About one hundred miners were left to work there after the initial excitement had subsided. Those not carried away with the quartz mania were hard at it getting their placers ready for production.

The enthusiastic promotion of the new gold and silver lodes at South Boise started early. The Elmore, thought from the first to be the richest, had a notable publicist in H.T.P. Comstock, who sought to enhance his interest in the property by pronouncing it to be fully as rich as the lode that had been named for him in Nevada. Rich outcrops lent support to Comstock's extravagant prophesy, and when arastra production commenced late that summer, South Boise quartz promoters had some high yields to talk about. Comstock's Elmore turned out seven tons at $347 a ton, and another property did still better with a total of $1,480 from only four tons of ore. Shares in the Ophir then sold at $25 a foot, whereas the Idaho, the original lode discovery, was valued highly enough to be "not for sale at any price." South Boise miners by that time looked forward with confidence to a rush of five to eight thousand newcomers in the spring of 1864, a misfortune from which the district happily was spared.

Placer mining accounted for by far the greatest part of South Boise gold produced in 1863. Before the season ended, several localities had contributed significantly to the region's yield. Besides the early activity of Red Warrior, Happy Camp — located on Elk Creek below the mouth of Bear Creek — was the scene of considerable effort during the summer. By fall thirty-five companies, ranging from one to five miners each, were hard at work, averaging $12 to $25 per man per day in Happy Camp alone. Some additional placering was in progress in Blakes Gulch, as well as on the lower Feather River in the area of the original South Boise discoveries that spring, on parts of Bear Creek which was regarded as rich at the time of original discovery, and perhaps to a limited extent in Hardscrabble Gulch on Elk Creek. Altogether, the South Boise mines had a population of 560 when the 1863 census was taken in September. South Boise was already as large as Pierce and Florence, and only slightly smaller than Warren and Silver City but larger than Elk City, Newsome, and the Salmon River camps. The one region which none of these remotely compared with in size and population was Boise

Basin, which contained Idaho's largest camps.

Surface prospecting of the South Boise mining region was adequate enough in the first season to disclose most of the better known mines, including the later big producers. Aside from the major properties near the Elmore, the Ophir on Elk Creek, the Bonaparte on Cayuse Creek, and a number of leads at Wide West Gulch on Red Warrior Creek had been discovered.

But the geologic structure of the Rocky Bar Basin was understood quite imperfectly at first. A confusing system of parallel veins, faulting transversely to the left, gave the early prospectors the notion that the mines ran generally north and south, and claims were taken up accordingly. When further examination disclosed that the claims ought to have run more like east and west, major adjustments were required.

By the fall of 1863, a total of "some 200 well-defined lodes" were thought to have been identified in the district. But values of these "well-defined lodes" at depth remained undetermined at the beginning of 1864. Development shafts had not been sunk by more than fifteen to twenty feet, and no one at that time could distinguish rich surface concentrations from veins that would continue to be productive when serious mining got under way. On top of that, arguments developed over the identity of new lodes. Claim disputes often arose when alleged discoveries might have been traceable to veins already taken up. The location of the Confederate Star on February 9, 1864, on a vein and claim thought by G.F. Settle to be the one he had found May 9, 1863, eventually led to a lawsuit won by the Confederate Star people. A number of litigations such as this soon plagued the district.

Complaints naturally were expressed against the indiscriminate promotion of surface pockets that looked for a time as if they might be rich ledges, of spurs of veins already included in known claims, of barren veins whose rich assays came from other mines, and even of known rich ledges. A cautious reporter protested on February 22, 1864, that all too many of the good lodes were located and sold more than once, and

> that not one in ten of the ledges in the Recorder's books . . . have yet been prospected enough to ascertain whether they are ledges at all, much less whether they contain gold. Hundreds of claims on all these ledges — both real and bogus — have been located and recorded, for which the claimants have not even troubled themselves to look; these are the cheap interests sold in rich ledges and new discoveries.

Even Comstock, whose arastra continued to grind out about $270 a ton from the Elmore, had to admit

that his new lode, while rich, was "spotted" in its values. But interests in the Elmore were still valued at $50 a foot, and the future seemed bright.

The difficulty in transporting equipment and supplies to the remote South Boise mines retarded development of the district severely. In the summer of 1864, Julius Newburg's South Boise Wagon Road Company began constructing a toll road to Rocky Bar. While the road was being built, the processing of high-grade quartz in primitive arastras expanded greatly. Arastras were made inexpensively from local materials and depended upon horse or water power, both of which were readily available in the region. By the end of the summer of 1864, the number of arastras had grown from ten to eighty, and the larger ones were capable of milling 1 to 1½ tons daily. With values ranging from $75 to $300 a ton in 1864, something like $130,000 to $160,000 came from the South Boise quartz mines that season.

High-grade ore had to be sorted out laboriously to supply the arastras, and recovery left much to be desired. The season's arastra production from the Elmore (the major South Boise lode, but not quite the main producer in 1864), from which only 100

Although they served other parts of Idaho, these water power arastras were typical of those at Rocky Bar

tons were milled in arastras in 1864, was $30,000 instead of the $50,000 that Dr. S.B. Farnham, agent for the Idaho Company, had estimated the ore to contain. H.T.P. Comstock was disturbed to recover $10 to the pan (an extremely high rate considering that a few cents to a pan would set off a gold rush) from tailings taken 100 feet below his arastra. The arastra process was so wasteful that operators were better off to ship some of the best ore out for testing and milling. Wilson Waddingham sent ¾ ton of South Boise Comstock ore all the way to San Francisco where $600 was recovered from the lot showing that he had at least $800-a-ton ore. He also obtained assays as high as $7,112 on the same South Boise Comstock, $5,589 on the Confederate Star, and $7,434 on the Elmore. These came from extremely high grade specimens that were characteristic of the district but not present in anything remotely like commercial quantities.

⚒   🐎   ⚒

The completion of Newburg's road to Rocky Bar on September 5, 1864, was an occasion for great rejoicing. Except for a steam sawmill (rated at fifteen horsepower and capable of turning out 4,500 board feet a day in July) that Cartee and Gates had packed into Rocky Bar in the spring of 1864, heavy equipment had waited for the road to be completed. By the time the road was finished, Cartee and Gates had a five-stamp custom mill set up near Rocky Bar. This small mill could handle five tons a day, compared with 1 or 1½ tons for a large arastra. Crushing 150 tons of $100 Elmore ore the first thing that fall, Cartee and Gates' mill suddenly increased the total production of the Elmore that season to $45,000.

Six more stamp mills were being brought in or were being erected during the fall of 1864. The most substantial of these, the twelve-stamp mill that the Idaho Company had freighted from St. Joseph, April 20, 1864, across the plains at a transportation cost of 30 cents a pound, or $8,400, reached Rocky Bar in November. Although complaints were already voiced against stock market manipulation on such notable properties as the Elmore, where the old Washoe "freezeout game" of letting a tunnel cave in to discourage stockholders so that the management could increase its interests at small expense, great profits were expected from the district as soon as stamp milling could get under way. After all, men had been making $20 a day just hand mortaring samples while prospecting the Ophir. Once a fast, efficient stamp recovery process could be installed, the mines were expected to prove their worth brilliantly.

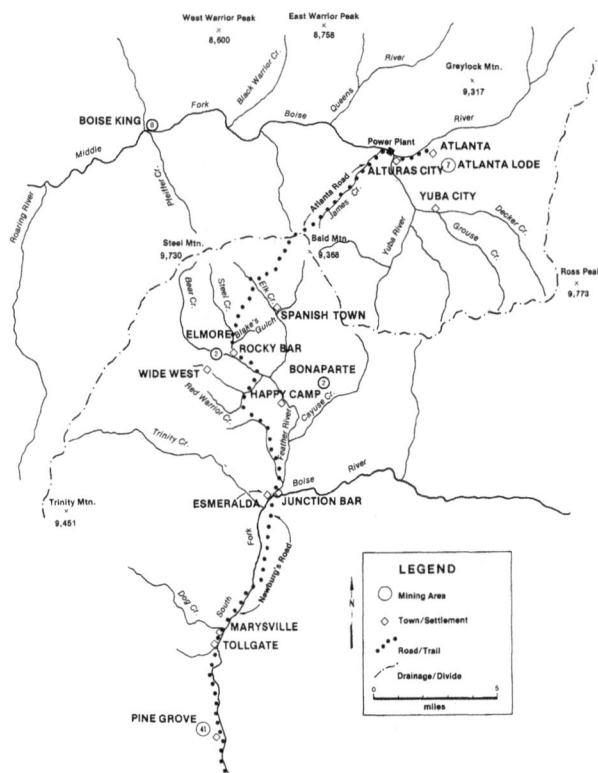

Rocky Bar and Atlanta region. Numbers in circles refer to index map of mining areas, page 2.

To insure that the stamp milling of South Boise gold and silver ores would be efficient, Wilson Waddingham and J.W. McBride took advantage of Newburg's road to haul another seven tons of ore from various mines at Rocky Bar out to Portland for testing. When the ore arrived there on November 27, they were prepared to send some of their large samples all the way to Swansea, Wales, if necessary to determine the best process for gold and silver recovery. Meanwhile, about half of the six hundred men who had been in South Boise that fall were ready to spend the winter preparing for a big season of stamp milling the next spring.

Except for erecting the buildings and doing the other work required for installing stamp mills, the South Boise miners left their properties idle. Development work to block out ore was not regarded as necessary then. Quartz miners simply worked down from the outcrops of the veins and hoped that ore sufficient to keep their mills busy would be available. Because arastras crushed ore slowly so that miners had little trouble in keeping sufficient ore supplies on hand, this process did not induce anyone to get far enough below the rich outcrops to notice whether the veins amounted to anything at depth. Promoters in at least one case "salted and sold a blank ledge" to one South Boise stamp-mill company, and not until the mill had almost reached the district did anyone notice the entire lack of ore. Great care was exercised to make certain that a milling and recovery process proper for South Boise ores was used. If anything like the same care had been devoted to making sure that each of the stamp mills had sufficient ore on hand to work, large-scale production might have been possible much earlier.

With arastra production suspended in 1865, the quartz yield that summer came from two stamp mills that had been brought to the district late in 1864. Cartee and Gates found ready customers even though they charged $25 a ton, and the Idaho Company's twelve-stamp mill also ran all summer, turning out $800 to $1,000 daily. The Pittsburgh and Idaho Gold and Silver Mining Company thought it profitable to invest $140,000 in purchasing the Idaho mine and mill. Wilson Waddingham, whose company was capitalized at $600,000, was busy investing recklessly in other mines so to consolidate enough property to justify a large stamp mill. With a paid up capital exceeding $400,000 in New York investment, Waddingham's New York Gold and Silver Mining Company did not face the problem of having to manage on insufficient resources. (Less adequately funded companies had to try to operate from current proceeds, usually with disastrous results.) Waddingham arranged to bring in an eighty-horsepower, forty-stamp mill at a cost of $100,000 or more. Freight costs from the Oregon Steam Navigation Company's dock at Wallula overland to Rocky Bar ran to $40,000 alone. To haul the mill machinery, Waddingham required forty-five "mammoth wagons." This great mill, intended for the Elmore, had a capacity for handling seventy-five to one hundred tons of ore a day. While it was on the way to Rocky Bar — a trip requiring all summer and fall — Waddingham purchased for $27,500 in gold James O'Neal's ten-stamp mill that was capable of processing sixteen tons of ore a day. Used on the Confederate Star, which Waddingham acquired for $15,000, this smaller mill turned out $60,000 by March 1866, more than meeting expectations and justifying its cost.

⚒   🐎   ⚒

Some of the other stamp-mill companies were less fortunate, however. Not anticipating the time and difficulty that would go into bringing expensive stamp mills from San Francisco or Chicago to this remote district, and not capitalized sufficiently to spend a year or two getting a mine ready to

Early Rocky Bar

produce, a number of companies began to get into serious financial trouble. Labor costs were high — $7 a day per man, compared with $6 in the less remote Owyhee mines and with $3.50 on the Comstock; in those years of hand drilling, labor costs amounted to the greatest part of the expense of mining, once a mill was procured. Adversities arising from the serious difficulties in getting a quartz mill into production began to plague South Boise as early as the summer of 1865. S.B. Farnham's New York and Idaho ten-stamp mill had barely begun to crush rock on August 13 before insufficient capital reserves and high operating costs were causing trouble. By fall, unpaid teamsters had imprisoned Farnham, and his crew had barricaded the mill pending back wage payments.

An illuminating example of inept mine administration by another New York company active in South Boise was to be found in the misadventures of the Victor concern, whose Red Warrior mill commenced operation shortly after Farnham's failure. The Victor operation may be traced back to the summer of 1863 when Thomas J. Gaffney discovered some Red Warrior quartz lodes. That winter, Gaffney had gone to San Francisco to obtain capital for developing his discovery, and there he met Francis O. Nelson, whose experience was primarily as a ship's captain. Nelson was also one of the very earliest stamp-mill operators in California and was highly regarded by the old Californians. Together they organized the Victor Gold and Silver Mining Company of California, to which Gaffney deeded twelve hundred feet in five of his claims.

Gaffney returned to South Boise in April to manage the property, and Nelson set out for New York in July to raise more capital. Nelson promoted well, although he never had seen the properties which the Victor Company owned. On December 5, 1864, six incorporators, including Nelson, organized another Victor Gold and Silver Mining Company, this one of New York, capitalized at a million dollars. Nelson was appointed manager, January 16, on the assurance that he could bring in a mill and get it running on a self-sustaining basis for not more than a $40,000 investment. He was not limited to $40,000, however, and his actual expenses, including $9,300 for a fifteen-stamp mill in San Francisco, amounted to $36,500 before the mill finally began operations on August 28, 1865.

Cleanups, in which accumulated gold was retorted from mercury, were held every Sunday for sixteen weeks into that fall and winter. The first three did not amount to much, since granite and low-grade ore was used to get the mill into operating condition. After the mill began to

produce, Nelson kept right on drawing upon the credit of his Victor Company in Portland. Before that source of funds was finally cut off after October 28, he had used an additional $16,000, presumably for operating expenses, which raised the company's capital investment in the venture to $52,500.

For what the money was spent, aside from the $9,300 for the stamp mill or what his production totaled, the company never managed to find out. Captain Nelson seems to have run the enterprise personally, entirely too much as he might have run a ship. He never sent in vouchers to the company to account for more than the cost of the mill at San Francisco, and whether he was drawing upon the company's credit for purposes other than the mine could not be ascertained. None of his employees knew how much was produced. Only the wildest guesses could be made from information that the company had gathered after sending another member from New York on October 7 to investigate. Pending a report, the directors decided on October 19 to bond the mine and mill for $50,000 and to cut off any more credit to Nelson until his accounts were straightened out. Some of his employees, unable to collect payments on drafts that Nelson had made to them after the crackdown, learned that the company no longer was honoring checks. At last, on December 1, Nelson finally sent $2,367 (out of one $5,000 weekly cleanup) overland to New York as the initial return to the company on its investment. He seems to have been unable to continue milling very long. After he could no longer pay operating expenses out of company capital, the mill shut down on December 20, 1865. The reason given in 1866 for the long-continued shutdown was the need for parts which could not be obtained in the winter. That explanation may have been correct, although there was probably more to the story than that. Nelson's method of handling the product of the Victor mill seemed to have been about as skillful as his method of conducting the company's financial affairs.

Isaac Thompson, who worked in the mill while it was running, described the system in some detail in an affidavit. In it, Thompson refers to himself as the "deponent":

> . . .the first three cleanings up were not of much account, because a good deal of granite, quartz &c. was crushed merely for the purpose of wearing down and smoothing the machinery and batteries. That the fourth and fifth cleaning up was good and that the subsequent cleanings up were very inferior, but deponent is unable to state the precise amount of yield. That on one Sunday a cleaning up yielded a wash basin full holding about one gallon and a half and four pint

bowls almost full of amalgam and that the weight of this amalgam so obtained must have been between seventy and eighty pounds; that another clean up yielded the same wash basin full and four sized tumblers full of amalgam and that the weight of this amalgam must have been from seventy to eighty pounds; that the deponent on one occasion saw in the bedroom of the house occupied by said Francis O. Nelson and family a wash pan about two-thirds full of amalgam containing by measure about eight quarts of amalgam. That deponent is unable to state what became of the proceeds of the different cleanings up or the precise amount of the yield of the same, as the said Francis O. Nelson kept all his business to himself.

Nelson's associates in the Victor Company, being unable to find out anything from Nelson or to determine in any other way what the mill had produced, concluded after an extended investigation that the cleanups ought to have averaged $4,000, and thus to have totaled $64,000. In assuming such a high average, they were almost certainly overly optimistic. Nelson found some witnesses to allege that the ore generally was poor or worthless, and his witnesses may have been right. In any event, the mine had failed, and Nelson was removed as superintendent on March 16, 1866. Whether he had put company money into his own projects, or whether he had applied the Victor funds in developing an unprofitable mine, cannot be ascertained. If he was an honest superintendent — and most likely he was — he certainly showed his utter incompetence in handling the company's financial accounts. And if he was an authentic swindler, the very least he could have done would have been to supply his company associates with some false accounts. Some such method, at least, is how William M. Tweed, one of the six organizers of the Victor and vice-president of the company, would have handled it, if his accomplishments in defrauding in New York City through the machinations of the notorious Tweed ring are any index at all. But compared with the Tweed scandals (not yet revealed in 1866), Nelson's defalcation in the Victor case is entirely unworthy of mention.

Tweed and his New York Victor associates had been clever or slippery enough to arrange things so that by the end of October 1865, when they had bonded the company's property with a $50,000 mortgage, their loss would be slight indeed. And after Nelson's one remittance was taken into account, they had salvaged $52,367 out of the $52,500 which Nelson had spent before they cut off his credit in Portland. So unless they lost from assuming some of Nelson's later obligations, in the end they were out only the cost of investigating

Nelson's incompetence and the mine's failure. Later in 1866, the mortgage holders foreclosed, and the Victor mill and property were auctioned at sheriff's sale. Thus, the Victor creditors and the mortgage holder assumed the main losses in this whole operation.

Understandably, Rocky Bar merchants resented such a method of financing unsuccessful lode operations. A South Boise promoter warned of hazards to local suppliers who might advance credit to distant companies in a letter from New York, March 25, 1867:

> The stockholders and directors of the N.Y. & Idaho and the Victor mining companies have resolved to worry out the creditors by protracted litigation. The members of these companies are men of wealth and can easily keep the suits in court for years. Judgement is not very swift or certain in New York, so I learn from attorneys here. From present appearances it seems as though the creditors have but a slim chance of contesting their suits through the courts of New York. A want of concert of action by the creditors in obtaining their judgements and prosecuting their claims has worked to their disadvantage. Let the mining communities of Idaho take warning and not trust these N.Y. companies for one dollar hereafter, for these New York rascals are worse by far than our own, and they live a long distance from Idaho.

⚒ 🐎 ⚒

In happy contrast to Farnham's and Nelson's disasters, several more mills began production by the spring of 1866. By then, South Boise had more stamp mill capacity than any other Idaho district. Rasey Biven's Wide West property milled $1,000 a day, and another company recoverd $5,000 in twelve days of February 1866. Wilson Waddingham did still better, with a $7,200 run of $82 ore in thirteen days, and by March, 1866, his Confederate Star, wisely keeping a stabilizing reserve of 150 tons, had ground out about $60,000 in four months.

When Waddingham put his ten-stamp mill into operation, he learned that he really did not need his $100,000 large forty-stamp mill at all. His small mill could handle both the Elmore and the Confederate Star. Mining from the Elmore (the major Rocky Bar lode) proved difficult. As soon as any depth was attained on the lode, his steam power plant had to be used to pump water from Bear Creek out of the Elmore, rather than to run a sawmill, as was planned. Twenty years went by before the Elmore could be developed profitably.

Meanwhile, Waddingham concealed his massive blunder. With a useless forty-stamp mill at

Rocky Bar, he began to invest in the enormous Atlanta lode. Here his $100,000 mill could be transported over a high ridge "at a trifling cost." That way, he managed to declare a 1½-cent dividend on $600,000 capitalization on December 1, 1865, and had a number of handsomely profitable (but relatively small) mill runs to report. With the best properties around Rocky Bar, Waddingham was recognized as a respectable, legitimate operator. At a Rocky Bar testimonial dinner early in December, "in variety and style never before seen in the Territory," Waddingham received a well-deserved tribute: "There is no doubt that to Waddingham's moral worth, strict business habits, and to his representations, the community of these regions are mainly indebted."

Waddingham himself spent the winter in New York with other South Boise agents of capital. But he found the mining market there badly depressed. New York investors had little way of distinguishing the legitimate mining companies in the West from the frauds. Because of failures, even with the serious companies such as Farnham's New York and Idaho, investors were becoming distrustful. Unfortunately for those reliable companies getting into production, the recent failures made investors fearful of putting up enough additional capital to meet the unexpected delays and higher costs involved in starting a mining operation. The series of failures of New York companies in South Boise continued after the announcement of the collapse of the Victor in May 1866.

A legacy of embarrassing debts, litigation, worthless stock, and adverse publicity afflicted the region with each failure. By June 2, 1866, more South Boise companies were in trouble, and even Wilson Waddingham's Confederate Star faced litigation for not paying dividends. Only two companies operated during the summer of 1866. Waddingham ran his mill through much of 1867, but he found it much more profitable to sell out his interest in the Elmore for $50,000 and to withdraw from the region. The Pittsburgh Company tried to sink a deep shaft to develop the Elmore but failed to figure out how to pump an abundance of water from Bear Creek out of the shaft; anything like large-scale development of the Elmore had to wait twenty-four years for British capital and more advanced technology.

Confidence in the future of the South Boise mines, as well as in the other Idaho quartz districts plagued by early stamp-milling disasters, survived undiminished by the setbacks to large-scale mining. No one doubted the richness of the mines, even of seemingly unproductive operations. That the mines had great potential was shown in the bitter, and

These photographs of the Idaho and Pittsburg mines were taken shortly after Fort Emmett blew up in 1866

sometimes violent, clashes over claim jumping or alleged claim jumping. Litigation between G.F. Settle's Idaho No. 2 and the Confederate Star had bedeviled the major quartz area around Rocky Bar through 1866. No sooner was the case settled on September 29 by a jury in favor of the Confederate Star, than another contest arose when claim jumpers built Fort Emmett on the Idaho lode. The Emmett Company mined the fort with quicksilver flasks of balls and powder, which were fused into the Emmett tunnel. Somehow the whole thing blew up on September 29, 1866. Fort Emmett was vacant at the time, and no one was injured when it was demolished. All of this tumult, though, revealed that the properties were not regarded as valueless.

Blame for the stamp-milling failures was usually put on the New York and other outside companies for mismanaging their enterprises and thereby casting aspersions on the integrity of the district. "Idaho has suffered many things from successive crops of knaves and fools who have dabbled in her mines; and the stockholders of the East have reaped a rich harvest of assessments and lawsuits in consequence of sending them here." That, at least, was the opinion of James S. Reynolds of the *Idaho Statesman*, September 10, 1867.

Naturally, there was more to the story than that. Many of the outside investors objected to having been beset by frauds and swindlers. And some of the other causes of difficulty, already indicated, were appreciated by them. Discussion of stamp-milling failures dwelt at the time around the South Boise disasters (the entire matter is considered on pages 85-86). Regardless of the explanations given for stamp-milling collapse, the conclusion was universally the same: the mines were good, and proper development would make them pay. In fairness, perhaps too much was expected of the mines initially, but in the end they did produce.

⚒   🐎   ⚒

For many years after 1866 and 1867, unpretentious arastra operations and a few modest stamp-milling enterprises were about all that survived the failure of early, large-scale quartz mining in South Boise. The placers, likewise, seemed by 1867 to be mainly suitable for Chinese operations. A small, cooperative company on Red Warrior was able to work economically and profitably that year. The five-stamp mill at the Bonaparte ran with some success in 1867 and 1868. In 1869, Rocky Bar was described as "dull and looking rather dilapidated very much in need of repair." Many arastras were still going, and in that respect, times seemed almost like the big days of 1864. But there was an

important difference. Early arastra operations were regarded as preliminary to large-scale stamp milling; by 1869, such operations were regarded as a substitute for unsuccessful stamp milling. Exceptions to the stamp-milling failures were few. By superior management and by working better grade ore, John McNally was able to keep his Wide West mine and mill on Red Warrior in operation through 1869. By the end of the year, his was the only stamp mill going.

Thus the failure of stamp milling in South Boise in 1866 and 1867 had proved to be a serious setback for the quartz mines there, though some compensations came with the failures. Expensive mills and equipment had been brought into the country and were available at rather low costs when auctioned at sheriffs' sales. In the lean years before railroad transportation, additional capital, and improved technology had brought big production to South Boise, some of these abandoned mills did much to tide over the mines which struggled along. Not until 1886 could the district be developed satisfactorily. In the interval, much that was done was called gouging, whereby miners unable to develop their properties in full took out some of the higher grade ore, which if anything, set back the mine because of the way they went about their work. During those years systematic mining and adequate recovery processes were neglected in favor of getting out what could be handled easily. Although all kinds of efforts, along with gouging, were made during the two decades to get big quartz mining enterprises successfully under way, unquestionably the initial South Boise gold rush and excitement had ended by the summer of 1866.

Many small operators, returning to arastras after stamp milling had failed, managed to maintain a modest level of production around Rocky Bar during the decade or more when gouging provided most of the mineral recovery. A revival of large-scale mining seemed possible after 1869, when the transcontinental railroad was completed across Nevada and Utah. The introduction of dynamite at this same time reduced the costs in hard rock drilling. Better hoists, engines, and pumps became available with technological progress. John McNally's well-managed mine on Wide West Gulch brought a substantial increase in production in 1870. The next year a Pittsburgh company introduced superior hoisting and pumping devices, which made possible the development of the Elmore mine under Bear Creek. Rather than spend excessive sums buying out the interests of the many owners of linear feet along the Elmore vein (because they owned claim footage instead of stock in a company), F.F. Oram invited anyone who held

Several important Rocky Bar mines (Elmore, Pittsburg, and Confederate Star) show in this picture taken before 1884

small segments to join his Pittsburgh Company in putting up development costs. After production got under way, these minor associates continued to participate in whatever profits — or losses — were realized. That way his pumping and hoisting service could handle the entire vein, rather than having several adjacent pumping plants operating on separate properties. A fifty-ton test run in 1872 yielded $4,000 from selected high-grade ore. This system would have worked still better if the superintendent had not sneaked off with the proceeds, leaving his miners unpaid and his participating associates with no return on their investments.

The next year a new manager of the Elmore mine succeeded in milling another forty-five-ton test run worth $5,000, after another pump was installed. This success led to considerable development in 1874. Trying to operate during the Panic of 1873 upon milling returns worked out poorly, both for the Pittsburgh Company and for Warren Hussey who employed the same system after taking over the Wide West from John McNally for $22,000 in 1874. Hussey had no way to continue production when his mill broke down, and he could no longer pay his miners. Trouble with the Pittsburgh's hoist foundation and with its recovery equipment forced the company to shut down in 1875 after sinking the

Elmore shaft to a depth of 225 feet. Gouging failed to work effectively for either of these major South Boise producers. When they contrived to produce gold, they naturally had to start paying returns to outside investors or lessors at a time when they lacked capital sufficient for effective mine development. Superintendents, who were paid a percentage of their production, had more incentive to gouge out a small amount of high-grade ore rather than develop their properties for large-scale mining. Aside from some small arastra operations, little lode production could be accomplished until after 1880.

The construction of the Oregon Short Line across southern Idaho in 1882 and 1883 eventually brought prosperity to lode districts such as Rocky Bar and Silver City, which also had to shut down in 1875. While the railway was being built, True W. Rollins, who had purchased from 1876 to 1879 much of the Elmore and associated properties with New York capital, got equipped to develop his mine up to a depth of fifteen hundred feet by 1882. The settlement of fifteen years of Idaho and Vishnu litigation in 1880 ended a wasteful era of leasing and gouging. In 1883, these companies finally managed serious production with a $100,000 yield. Then they got back into litigation, and Rollins

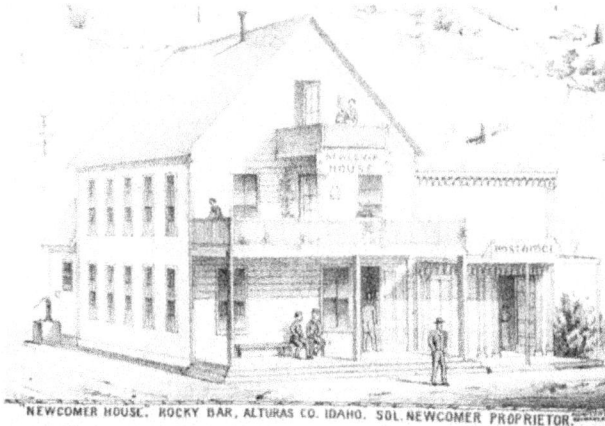

NEWCOMER HOUSE, ROCKY BAR, ALTURAS CO. IDAHO. SOL. NEWCOMER PROPRIETOR.

Alturas Hotel, Rocky Bar, 1890

(after investing $150,000 in developing the Elmore) found that he could not operate after all.

Finally British capitalists acquired Rollins' property along with other important Rocky Bar mines aside from the Vishnu. By completing a fifty-stamp mill on November 16, 1886, they were equipped to operate the Elmore efficiently. Their initial year returned a profit of $320,000 out of a $460,000 yield. They continued a steady production with the best modern equipment until March 5, 1889. After a long, expensive effort at developing more ore, they gave up in May 1892, having sunk their shaft seven hundred feet to prove that Rocky Bar did not have good ore at depth.

By 1892, other companies had also realized most of their production. Limited mining continued for another half century. Yet most of Rocky Bar's mineral yield came in a short period after 1886, following more than two decades of effort to solve problems of mining. If British investors had known to stop production in the spring of 1889, their mines would have shown an acceptable return. But as their enterprise finally worked out, they learned more than they really wanted to know about the

British investment financed construction of this fifty-stamp mill in 1886 (photograph taken in 1890)

lack of ore in the lower levels of the Elmore.

Production around Rocky Bar did not end entirely in 1892. Although a disastrous fire on September 1 left over half of Rocky Bar's 200 to 300 residents homeless, reconstruction provided a new town that lasted for more than another generation. About thirty Elmore and Vishnu miners continued to explore those properties until 1896, when a firm from Scotland undertook a bedrock flume project on Bear Creek to recover amalgam lost from earlier stamp mills. Over $40,000 (about half enough) was invested in this project, which included 2,200 feet of constructed flume and a steam derrick to remove boulders. Additional funds were needed to bring water from the upper part of Roaring River over a high ridge to Red Warrior. This overly ambitious project could not be completed, and Rocky Bar declined still more. Another scheme to start up a twenty-stamp Bonaparte mill failed to accomplish much in 1904. But Junction Bar placers were tested with favorable results in 1906, and some unsuccessful efforts at Bonaparte (which claimed a previous $600,000 production), at Elmore (with a $3,000,000 record), and at Vishnu followed a year later. Finally, a stationary dredge at Feather River, powered by a 175-foot head of water from Cayuse Creek, was used from 1910 to 1915. Eventually a standard floating dredge commenced production there on August 21, 1922. By 1927, 33,000 ounces of gold came from that operation, which required an initial $500,000 investment in equipment.

Low operating costs after 1929 encouraged a number of modest efforts to reopen Rocky Bar mines before wartime restrictions forced all gold mining companies to suspend work in 1942. Aside from an Ophir promotion that led to a $17 sheriff's sale of that old property, which had only an $80,000 production record from a vein that looked like a major lode, another forty years of inactivity followed, and Rocky Bar almost disappeared. Then in 1982 mining resumed right on the townsite of Rocky Bar. A large backhoe and loader operation, capable of handling a thousand yards a day, was employed to overcome a previously unmanageable problem of moving large boulders (for which Rocky Bar is named) so that the deep placers could be mined.

OWYHEE

Silver mining in the United States, except for a few old Spanish workings in Arizona, had not much more than started when mineral development of the Owyhee country gave Idaho an important share of the national output of that precious metal. Although Owyhee began as a gold-producing region, miners soon found silver to be their major resource. Within two years of the rush to Owyhee, the richness of the new region had impressed the nation in a way that none of the earlier southern Idaho gold discoveries could match.

Albert D. Richardson, a noted newspaper correspondent who traveled all over the West, concluded after personal inspection that Owyhee had the "most abundant gold and silver-bearing rock ever found in the United States." Attention in the early years focused upon the celebrated War Eagle Mountain mines, which Richardson described as constituting "the richest and most wonderful deposit of quartz yet discovered in the United States, even eclipsing the famed Comstock Lode." The facts then known gave support to Richardson's claims. Although the War Eagle veins

were extremely narrow, they tended to widen at depth, and their richness surpassed anything else yet found in the nation. Later discoveries and development of far more massive orebodies in the Comstock put that pioneer silver district far ahead of Owyhee in total production; but in the earliest years of Silver City, the War Eagle Mountain mines had richer ores and a larger initial production, partly because mines at Owyhee were able to take advantage of Comstock recovery methods, which were developed just in time for use in Owyhee. Furthermore, because the extent of the early orebodies had not yet been determined for either district, these miners had grounds for optimism that theirs was indeed the superior one.

By putting technical processes developed on the Comstock to use in Owyhee, the miners were able to get quartz mining off to a more successful start than in the other quartz districts in Idaho. Stamp milling of Owyhee ores went through much of the same difficult pattern that retarded development of lode properties in Boise Basin and South Boise. But Owyhee had mines so rich at depth that some of them were able to operate for a decade or so before their initial collapse. The general pattern of development of Owyhee quartz mining was much like that of the other Idaho districts. Many similarities between Owyhee and South Boise may be noted, especially through 1866.

D.H. Fogus of the Boise Basin discovery party panned gold in the Owyhee country on June 28, 1862, and an emigrant caravan headed down the Snake River found gold deposits on Sinker Creek and high up Reynolds Creek on September 6-7. Serious interest in the region was shown the next year. Then Michael Jordan's twenty-nine-member party went out to make a more thorough inspection.

EXAMINING THE LEDGES ON WAR EAGLE MOUNTAIN.

INTERIOR OF A QUARTZ MILL.

These two illustrations come from Albert D. Richardson's segment on Owyhee in his travel volume, *Beyond the Mississippi*, published in 1867.

Going over to Fort Boise of the Hudson's Bay Company at the mouth of the Boise River, the crew crossed the Snake River there and proceeded eastward on the Emigrant Road, a variant of the Oregon Trail, that ran along the south side of the Snake.

On this route, they were following Moses Splawn's course of the year before when he and his party, after a vain search for the Blue Bucket mines, had come that way from Owyhee to Boise Basin where they did find gold. O.H. Purdy, one of the Jordan party, later explained that the Jordan group was also out looking for the mythical Blue Bucket placers, and the course of the men along the Emigrant Road as far as Reynolds Creek suggests that possibility, except that they were on the wrong trail. The Blue Bucket party had come through Boise Valley, rather than along the south side of the Snake River. Furthermore, the alleged Blue Bucket location was somewhere in central Oregon west of Fort Boise, not east of it. So if the Jordan group was at all serious about the Blue Bucket hunt, the men were looking on the wrong road in the wrong part of the country. By 1863 though, a great many prospective ventures were represented as Blue Bucket searches. Whether the men out looking were serious about that possibility or not, Sinker Creek in the Owyhee country got its name from the Blue Bucket craze. Leaving the Emigrant Road at Reynolds Creek, Jordan's party crossed over to Jordan Creek. There on May 18 at Discovery Bar, above what was later to be DeLamar, a '49er who was there reported:

> One of our inquisitive spirits carelessly scooped up a shovelful of gravel and 'panning it out'—found about a hundred 'colors'. . .Picks and shovels were wielded with telling strokes among the slumbering rocks, gravel and soil. When near bed-rock, was seen, in pleasing quantities, the idol of avarice, the master of men, and the seductive and winning creature of women—GOLD.

Then, dividing up into two- or three-man squads, the twenty-nine prospectors spent the next ten days examining about fifteen miles of the stream. Satisfied with the results, they organized a miners' meeting, adopted laws, and opened up a mining region that proved to be as incredibly rich in silver as the Blue Bucket was supposed to have been in gold.

Proceeding from Discovery Bar on Jordan Creek, the site of the original Owyhee mineral find not far above later DeLamar, the Jordan party worked its way up to better placers about two miles below the townsite soon to be known as Silver City. There they established Happy Camp, organized the mining district, and began to take up claims.

Perhaps they were too energetic in taking up claims. New comers complained that each of the twenty-nine original members had three claims, a total of nine hundred feet apiece. Although the Jordan Creek placers extended from four to ten miles, most of the better ground was appropriated by the original Jordan party, before the rush to Owyhee could even get under way.

When word of the new mines reached Boise Basin, and the returning discovery party prepared to set out again to work the new Owyhee placers, hundreds of others followed them to the new district. Because of the limited ground available, "a great many were disappointed, and some hardly got off their horses, but joined with the balance of the disaffected new comers to curse the country, the camp, and the party that found it." Returning to Boise Basin with tales that Owyhee was nothing but a humbug, these unhappy prospectors not only discouraged others from setting out for the new camp, but their howls of anguish also diverted suppliers from rushing in with equipment. During the initial season, Owyhee suffered considerably from the lack of tools and mining gear. Shortages of supplies held back development that first season, and a shortage of labor forced wages in Owyhee up to $8 a day. Shallow and easily worked, the new Jordan Creek placers yielded $15 to $18 dollars a day during the initial three- to four-month season while the water lasted. Gold-bearing gravel could

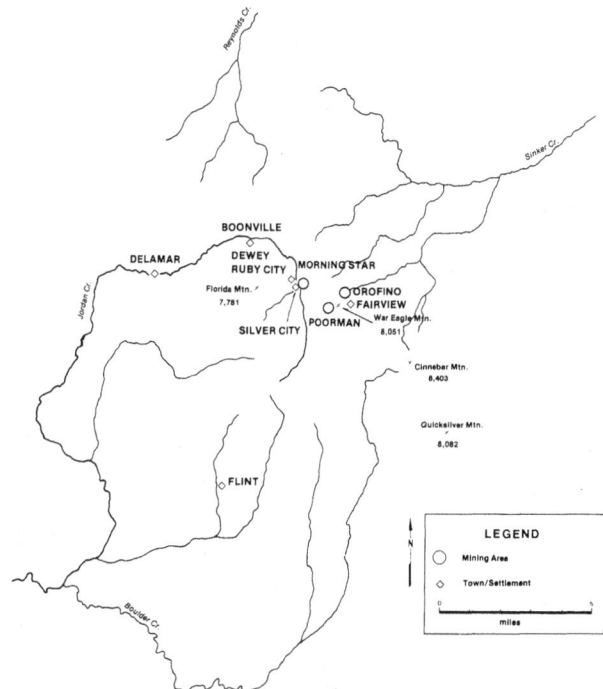

Silver City, DeLamar, and Flint mining area

be shoveled directly into sluices, since bedrock was only three to six feet below the surface. Taking everything into account, the miners who actually had claims in Owyhee did well enough. In spite of high costs, most managed to clear from $300 to $1,000 over expenses before water ran out and the first season halted.

⚒ 🐎 ⚒

If the Owyhee mines had contained nothing more than the Jordan Creek placers, the district would not have gained much prominence. Except for a few hydraulic giants and some Chinese operations, most of the placers were largely finished in the 1864 season. But two months after the original discovery, R.H. Wade of the Jordan party found a quartz lode in Whisky Gulch not far above the later site of Silver City. Not long after that, a group of prospectors on the other side of War Eagle Mountain, tracing some placers in the head of Sinker Creek, followed the gold up to veins on either side of the draw they were working on. A few weeks later on August 15, 1863, they located their new quartz discovery as the Oro Fino ledge. These veins, on the surface at least, were promising gold prospects. The Oro Fino, it was noted, outcropped "with gold crusting the surface of the rock in a most enticing manner." Although the Jordan Creek placers were an alloy of roughly half gold and half silver, successful miners leaving the district as late as October 6, 1863, noted that as yet the lodes were all gold with "no signs of silver."

Just as the exodus of Owyhee placer miners took most of the population away, another lode discovery, this time primarily silver, was located on October 14 not far from Jordan Creek below

Oro Fino mine development in 1866

Whisky Gulch. This silver property, called the Morning Star, was on the other side of War Eagle Mountain from the earlier Oro Fino. For the moment, most of the placer miners paid little attention to these developments. "But as winter approached, away they went, like blackbirds, to a more congenial clime, to spend the winter and their money, and in the spring [1864] returned as poor as crows; and who could otherwise expect but that an epidemic of hard times would scourge the camp?"

Late in the fall, speculators perceived the value of these new quartz properties and quietly began to buy claims for $2 to $3 a foot, without mentioning the possible value of the new lodes. By December, fabulous assays began to stir up interest. E.T. Beatty, a leading citizen of Rocky Bar, visited Owyhee in November and returned with average specimens of Morning Star rock running $2,800 in gold and $7,000 in silver to the ton. These assays far surpassed those of the Comstock. Owyhee now joined South Boise as a quartz area of unbelievable possibilities.

Astonishing quartz prospects promised Owyhee a permanence that the Jordan Creek placers could not possibly have held. Boonville, with thirteen houses, had supplanted Happy Camp as the placer mining center; and with the Morning Star discovery, Ruby City was planned and to some extent built in November. Still another town, later to be named Silver City, was laid out almost a mile above the new community of Ruby City. But since snow covered the site, no houses were started there that winter. Boonville and Ruby City had populations about 250 each by February 1864, although conditions were primitive as yet. The *Boise News* on February 20, 1864, mentioned that Boonville "is located on Jordan Creek at the mouth of a small gulch, and between high and rugged hills. Its streets are narrow, crooked and muddy, resembling the tracks or courses taken by a lot of angleworms, and might have been laid out by a blind cow to look as well." By then, two new quartz lodes were reported to exist near Boonville, "both considered good—if they ever find the ledges."

Preparations for extensive construction were under way even though prospecting or development of quartz lodes was not possible during the winter. To build the towns and prepare for the next season, fifty-six whipsaws were supplying lumber at $40 a hundred feet. Shakes commanded $6 a hundred. With such high prices and prospects for a still greater construction boom in the spring, the population of Owyhee had risen to about one thousand by February 1864.

Actual mining in the winter was limited to some hand mortaring of quartz outcrops. No more could be done. But even that slow means of working samples was profitable; so handled, both the Oro Fino and the Morning Star were good for $20 to $30 a day. J. Marion More, leader of the party which had founded Idaho City in the fall of 1862, had important interests in both of these properties, and on February 9, 1864, he attracted wide publicity by recovering nine ounces of gold and silver, valued at $13 a pound, from only 1¼ pounds of rock. Then, in Portland on March 15, professional assayers reduced a batch of Owyhee ore at the rate of $3,000 a ton. Profitable hand mortaring continued into the spring. D.H. Fogus' company produced $200 from another new discovery in one day, May 21. All kinds of veins were being located on War Eagle Mountain between the Oro Fino on the east side and the Morning Star on the west. The importation of stamp mills to work the new veins was only a matter of time.

⚒   🐴   ⚒

The development of even the best of the new Owyhee lodes had not proceeded appreciably before arrangements were made to bring in stamp mills. By May 23, 1864, the Oro Fino shaft was down only eighty feet, which was the deepest so far. In that distance fortunately, the vein had widened from six inches on the surface to thirty inches at depth. D.H. Fogus had started to run a tunnel on it, as well as to build a mill. Ore was also being extracted from the newer Morning Star, although it had been prospected only to depth of twenty feet. Both of these important properties had come for the most part under the control of J. Marion More and D.H. Fogus, and until the summer of 1866, most of the development and production of the Owyhee region was the work of More and Fogus. Each had done well in Boise Basin, and now they were investing some of their gains by acquiring the most valuable earlier Owyhee lodes and bringing them into production. They had to go to some expense, for the Oro Fino was selling at $100 a foot in the spring of 1864. Furthermore, bringing in a 10-stamp mill could cost around $70,000. Such an investment, however, might easily be recovered in only a few months operation — or so it seemed in the spring of 1864.

Aside from More and Fogus, a group of Portland investors, who had profited handsomely from the Idaho rush, held important Owyhee interests. These were the owners of the Oregon Steam Navigation Company, holder of the Columbia River steamboat transportation monopoly, which had hauled thousands of miners to Idaho after 1860. S.G. Reed, who later owned the Bunker Hill and

Sullivan for a time and eventually endowed Reed College, and J.C. Ainsworth, who personally inspected the new district, thought highly of Owyhee quartz. Ainsworth brought in a superintendent who had operated widely in South America and Mexico, and lately on the Comstock. Ainsworth's description of Owyhee quartz had sounded preposterous to the experienced superintendent he had found. But after examining the properties for himself, Ainsworth's expert manager was convinced.

The Ainsworth water power mill, built on Sinker Creek three miles below the mine it was to serve, got into operation in July 1864, after a force of men had built a road to the newly installed ten-stamp mill. Erected at a cost of $70,000, this mill yielded $90,000 in its first forty-five working days. That initial operation, however, suffered severely from lack of water to run the turbine. Until converted to steam in the following October, the Ainsworth mill ran only intermittently. The original mill had been designed only to recover gold. Upon noticing that the ore soon turned largely to silver, the managers had to convert to a new process the next spring. Fortunately for them, Owyhee silver ore, while not free milling, was well adapted to the Washoe process that had been developed on the Comstock for recovering gold and silver ores, which were, it turned out, similar to those of Owyhee.

While the Oregon Steam Navigation Company miners were solving some of the problems in operating an Owyhee stamp mill, several other important developments were taking place. A mass of ledges, some three hundred in all, on War Eagle

AINSWORTH MILL Cos MILL, 10 STAMPS. 5 MILES E of SILVER CITY.

Mountain between the Oro Fino to the east and the Morning Star to the west were being located. But without development, no one could tell which were parts of the same vein or which had any great value. With too many prospects for available capital to develop, a great mania to bring in New York money swept the district.

One of the most ambitious of these projects was that of the Morning Star Gold and Silver Tunnel Company of Idaho, organized on August 15, 1864, with a $5 million capital stock (to be subscribed in New York) to run a tunnel similar to the Sutro tunnel project for the Comstock. For this project, the tunnel was planned to run twelve thousand feet from near the Morning Star right through War Eagle Mountain to a point near the Oro Fino. Aside from prospecting a host of intervening veins — or supposed veins, at least — this ambitious tunnel project was designed to develop some important properties at depth. Backed as it was by Governor A.C. Gibbs of Oregon and managed by G.C. Robbins and A.P. Minear (who were two of the most important Owyhee mining and milling superintendents), the enterprise seemed sound enough.

The only trouble was that the $5 million was not forthcoming. The project got a lot of attention in New York, but was not even heard of in Owyhee. Had something like this tunnel been driven, War Eagle Mountain might have been developed on a much larger scale, with great gains in economy and efficiency. One of the most important obstacles to mining the rich War Eagle lodes proved to be the failure to consolidate valuable and important adjacent properties on the same vein. At the Oro Fino, in particular, several adjacent, independent mines were worked at a substantial increase in cost because they were not handled in a single operation as they should have been.

⚒   🐴   ⚒

Times began to change again in Owyhee by the end of August 1864. The great quartz rush was over for the season, although other excitements were to follow in later years. A Ruby City correspondent suggested on August 28 that earlier in the summer Owyhee had been overrun with "an eager and expectant crowd, who attracted by the golden

Silver City's Wells Fargo office was under construction when this photograph was taken in May 1866

stories of Idaho, sought to better their fortunes, but they did not find gold on the surface; and now they wind their way homeward, cursing the country." Those who remained, though, were a sober and industrious lot:

> To a stranger our camp presents but a dull and dreary appearance. The usual appendages of every new mining camp — bar-rooms filled with a crowd whose besotted countenances tell us but too plainly that it is not to them we must look for advancement in a country's prosperity, and women, who, in the glaring impotence of dress advertised the avocation they pursue — are not to be seen here.

By now, A.P. Minear's custom mill was almost ready to start, and More and Fogus were beginning to get some real development work done on the Oro Fino and Morning Star. With the idle prospectors largely gone, and the rest settling down to work in earnest, Owyhee was about to establish a reputation that everyone hoped would bring in the capital needed to make the district really boom. Morton M. McCarver, one of the prominent Oregon capitalists interested in mining possibilities, became an agent for a substantial number of holders of undeveloped claims before leaving Ruby City on October 10, 1864, to spend the winter selling them in New York.

Glowing reports of a rich fifteen hundred-pound sample that Thomas Donovan had taken to San Francisco for testing gave the new milling enterprises grounds for optimism. If only they could get their mills to work, they would soon recover their investments. Donovan was well repaid for his trip. His three-quarter-ton load yielded over $5,000, assaying $6,706.61 in silver and $489.12 in gold, for a total of $7,196.73 a ton. Owyhee promoters pointed out that this ran nearly a thousand dollars above the richest Gould and Curry assays on the Comstock — a point in which they could take genuine pride, since the Gould and Curry was then the big attraction at Virginia City.

At the same time, another ton of Donovan's rock was milled in Wales for $6,500. (In those years, shipping rich Comstock ore to the efficient Swansea Mill in Wales was regular practice; only the less valuable of the Gould and Curry ore was processed in Virginia City, where only an eighty percent recovery was guaranteed. Swansea, in contrast, guaranteed to recover full assay value.) In Owyhee, milling was not quite as efficient in the fall of 1864. A.P. Minear's mill, its ten 450-pound stamps having an eleven-inch rise, started early in September. In the beginning, only some of the stamps were in use. By the end of September, the new mill operators were still trying to finish the initial batch of forty tons of rock. The first ten tons had produced two

This view (c. 1866) of the More and Fogus mill includes a Silver City photograph gallery

bricks of 12½ and 13 pounds that were predominately silver. The rest continued to yield handsomely, but at a milling cost of $100 a ton. Minear was having a hard time finding anyone to employ in the services of his new plant.

More and Fogus, whose mill began to operate on October 3, made out better. With a substantial enough development of their own properties on the Oro Fino and the Morning Star, they had plenty of good ore to keep operating. In fact, their ore was rich enough that $100,000 in rock was hoisted from their shafts by man-power at a profit. The tailings had to be saved, however, since the mill was not equipped yet to recover all the silver. In its first eight-day run on second-rate Oro Fino quartz, four hundred pounds-troy of gold and silver amalgam valued about $5 an ounce totaled $24,000. Then ten tons of Morning Star ore brought them some $25,000 in two days. On November 27, J. Marion More arrived in Portland with $60,000 in silver bricks. He had good reason for satisfaction with his operation. He and Fogus had employed twenty to thirty men all summer in building roads, sinking shafts, driving tunnels, and putting in the best stamp mill in the district. Furthermore, they expected to add twenty-four stamps the next summer to the ten they already had, and their energetic development served to keep their mill, and all the custom mills as well, supplied in ore right along. By December 1864, their Morning Star shaft was down 115 feet, at which point they found the vein increasingly rich while drifting along it in either direction from the shaft. They had to convert the mill to the Washoe process to complete the recovery of silver. But by keeping up a steady operation, they produced about a million dollars in the first year of milling which had commenced in the fall of 1864.

All three Owyhee mills were running by December 1864, and the country was booming. That winter, however, turned out to be the severest in years. The new mining towns were hardly equipped for the ordeal. In mid-November a Ruby City correspondent had this complaint:

> The houses in this place were evidently built for summer use, without any calculation on the storms and snow of winter. A stranger coming into camp to-day would have thought snow and rain-water were articles of traffic, as merchants had pans, buckets, and every available vessel employed to catch that which poured through from divers cracks in the roof.

Owyhee promoters who left the mines to spend the winter in New York peddling ledges of all sorts—good, poor, or nonexistent—met a fine reception. Idaho properties had an excellent reputation. In reviewing activity for the year 1864, the *Mining and Scientific Press* reported:

> Perhaps the most noticeable mining development of the past year, upon this coast, has been that of Idaho, embracing the three districts of the Boise Basin, South Boise, and Owyhee. . . . Some thirty arastras were at work in South Boise last summer. There are now some eight or ten mills, either running or about to start, at Owyhee and Boise Basin, and several others upon which the work of construction has commenced. A very considerable amount of bullion may be expected from Idaho another season. It is thought that the quartz mines there will compare favorably with those of any region yet opened.

A whole string of fabulous reports—such as one from Portland in *The Oregonian* of February 26, 1865, that even with imperfect milling, ten tons of Owyhee ore had produced $120,000—continued to

NEW YORK & OWYHEE G & S MINING CO'S MILL 20 STAMPS

impress outside investors. By April 21, 1865, J.W. Ladd wrote from New York to S.G. Reed: "There are many persons turning their eyes towards Idaho and if you have any interests there in mines they quite likely will be salable whether they are of any account or not after awhile." And still a little later, even after some prominent investors had been swindled on Owyhee quartz sales, the *Mining and Scientific Press* was unaware of the sharp promotional practices: "There is good reason to believe that Idaho is destined to become a most important and permanent mining region. Thus far operations there appear to have been conducted upon a sound basis with very little of the speculative features so characteristic of new mining localities." Local people in Idaho felt that the sale of small, unprospected surface seams of no discernible value to Portland, San Francisco, or New York investors would accomplish nothing for the territory as a whole, aside from giving Idaho a bad reputation. Complaints against such promotional practices began to come from Owyhee as well as from Boise Basin and Rocky Bar. Before an adverse reaction had a chance to set in, however, the possibilities for Idaho mining investment had become so numerous as to saturate the New York market, and by the summer of 1865 they were no longer "all salable at once."

All during the hard winter of 1864 and 1865, More and Fogus had their mill running, and kept pushing their development tunnel on the Morning Star, drilling night and day. By mid-June, their Oro Fino tunnel was reported to be in 500 feet, where it connected to an intersecting shaft 150 feet deep. The Morning Star shaft had been sunk 125 feet, at which point they had drifted 250 feet in good ore along the vein. Blocking out three thousand tons of Morning Star ore, which they expected to be sufficient to keep a sixteen-stamp mill busy for a year, More and Fogus anticipated a million-dollar recovery from that small quantity of rock alone.

By the summer of 1865 several new mills were being rushed in as a result of eastern investment. One of them was lost on the ship, *Brother Jonathan,* which sank off Crescent City on July 30, 1865. And another, belonging to John Shoenbar, was delayed considerably when the teamsters freighting it from Red Bluff could not find the road and lost their way for two weeks or so around Goose Lake. Several others were headed for Owyhee. None of these mills could get into operation before late fall or winter, but by the time they were completed, there was every reason to expect More and Fogus alone to have enough good ore ready to keep them busy.

Silver City boomed with the construction of mills. Each of these enterprises employed some

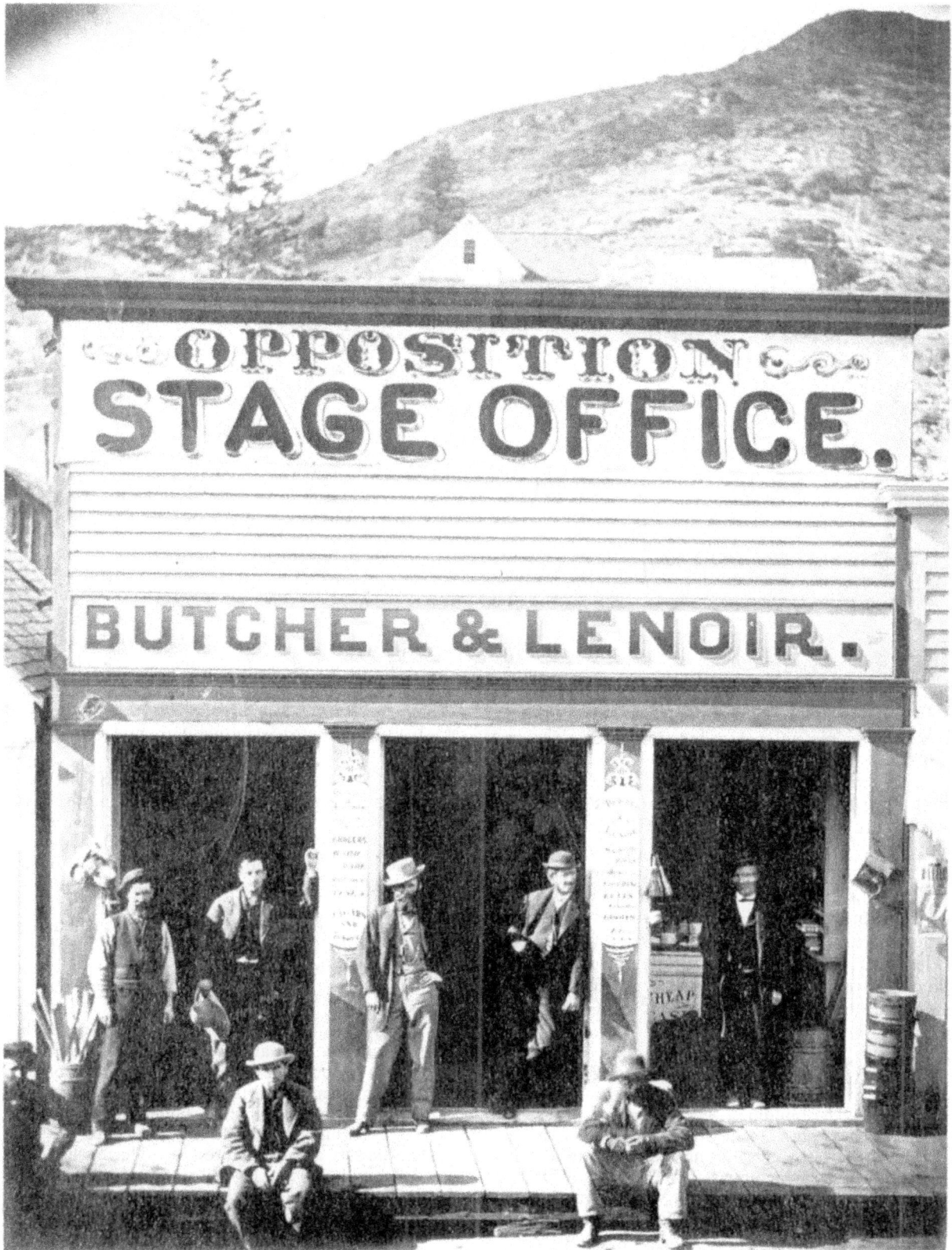

Silver City stage office in 1868

Silver City occupies a flat area along Jordan Creek below War Eagle Mountain. This photograph was taken in the summer of 1868.

thirty or forty men, and since the ones on Jordan Creek were located near Silver City rather than Ruby City, Silver City quickly emerged as the important center for Owyhee. The new mills not only lent more stability to the community but also had the appearance of permanence, suggestive of the long future which awaited quartz mining there. A description of John Shoenbar's ten-stamp mill under construction in the summer of 1865 suggests the amount of careful work that went into these enterprises:

> The site of the mill, about one-third of a mile above Silver City, is well chosen for that purpose. The building covers a space sixty feet deep and seventy feet in front, inclusive of the wings — the latter constituting apartments for the boiler and assaying apparatus. The walls of the building are composed of rough-hewn granite, such as is found in the immediate vicinity, and brick — the latter material being used principally on the corners, and around the doors and windows; the brick-work being painted a bright

red, the contrast afforded is of the most pleasing character. The different styles of architecture represented in the make of the various windows, add much to the attractive qualities of the structure. The finish of the roof, in the way of a cupola of lattice-work order, is also worthy of special mention; and altogether, the building surpasses, for elegance, strength, and durability, anything in the way of edifices we have seen in the Territory. We are informed by the proprietor, that a building composed of stone and brick, as this one is, costs but little more than a wooden one, and is far preferable. The mill now in course of completion, contains a battery of ten stamps, but the building is so constructed as to admit of another boiler the size of the present one, and ten additional stamps, with the necessary amalgamatores, & c. The machinery of the mill is of San Francisco manufacture, and of the first order. The framework around and above the battery, is all iron; the blocks on which the iron battery rests, are embedded in the natural granite, giving this part of the mill the appearance of more than ordinary durability.

Avalance office in Silver City, June 1866

Lincoln mill at Silver City in 1866

Packing in the machinery for several large mills of this kind employed a goodly number of mule trains that summer. These trains were described as having "more of the appearance of a cavalcade of mountain howitzers than anything else."

⚒   🐴   ⚒

Prospecting on War Eagle Mountain continued while More and Fogus ground ore and the New York companies built mills. One of a multitude of these new leads, the Hays and Ray, located on August 5, 1865, assumed great importance in mid-September. Altogether, eight 200-foot claims made up the Hays and Ray property. Somewhat small and not exceptionally rich at the discovery point, the vein suddenly became the most prominent of the War Eagle lodes. From a confusing set of facts, concealed at the time by parties engaged in expensive litigation, a mining engineer put together an accurate and clear account, the best that is available. It appeared as part of the Idaho report for J. Ross Browne's 1867 volume on the mineral resources of the United States:

> While the discoverers were developing their veins, a prospector named Peck found some very rich float-rock about 1,000 feet south of their shaft, and out of sight from its entrance. By a small amount of digging he reached the vein,

which he carefully covered with earth. Gathering up and secreting every piece of float he could find, he went where the discoverers of the Hays & Ray were at work, and after 'talking around' asked them where their claim was located and how far it extended in each direction. They showed him their boundaries, and walked directly over the spot where Peck had buried the vein, and at such distance beyond that he was convinced the claim embraced the rich ground. Peck continued to prospect in that vicinity, and cautiously commenced negotiation for the purchase of the mine. Not being satisfied with their figures, and there being few or no prospectors in the neighborhood, he left for a few days, thinking his absence would cause the owners to come down in their price. Before he returned another company of prospectors found the same spot discovered by Peck, called it the Poorman, and took out silver ore of great richness.

D.C. Bryan, the rediscoverer of Peck's find — now the Poorman, on September 13, 1865, located seven 200-foot claims in company with his associates, and recorded them two days later. Hays and Ray claimed the ground, but their vein was not uncovered or traced to the new opening. The Poorman Company refused to leave, and the Hays and Ray party had no money to pay for provisions or tools while they were tracing the vein. They gave Charles S. Peck a share in it for tracing it from their opening into the Poorman. Although the Poorman claims overlapped the Hays and Ray claims, the Poorman theory was that the two veins were separate and parallel. Should that have proved to be the case, each company would have been entitled to work its own vein. But the Hays and Ray locators intended to show, through Peck's tracing of their vein, that the Poorman was simply part of the Hays and Ray vein, with the exceedingly rich pay streak inside the boundary of the Hays and Ray claims. If they could prove that, the Poorman associates would lose the best portion of their new

War Eagle Mountain rises above Silver City. Poorman mine workings are to the upper right, outside this view.

discovery.

Before Peck could expose the Hays and Ray vein to the vicinity of the new rich deposit, the Poorman people quickly took out what ore they could. They managed to get in only six days work before Peck got so near to their deposit that the Hays and Ray combine had enough evidence to substantiate legal action for an injunction to restrain the Poorman. The small quantity of ore that they managed to get away with in that six days, however, turned out to be worth about $500,000. "In the history of silver mining," reported the Portland *Oregonian*, "neither Mexico nor Peru furnishes anything which can be successfully compared with it." Although the celebrated Poorman vein was only eighteen inches wide on the surface, it assayed eighty percent gold and silver, mainly silver. J.S. Reynolds, who reported the Poorman controversy at the time, remarked that "seriously there is no use describing the richness of the Poorman. No one will believe

without seeing it. Litigation, however, must greatly reduce its value, as there are several other suits in contemplation."

Aware that they were almost certainly on the Hays and Ray vein, the Poorman owners combined with the Oregon Steam Navigation Company's Owyhee interests to get financial backing for the legal battle that could not be avoided unless they simply gave up and left. Furthermore, they decided their only chance to salvage anything out of the Poorman was to rush what ore they could to the Ainsworth mill and to stop Peck from tracing the Hays and Ray vein right up to their operations.

The confrontation came September 24. Armed with shotguns and six-shooters, the Poorman defenders held back the Hays and Ray forces adjacent to the rich part of the Poorman: "The respective belligerents occupied a line about 100 feet long, and with a brush and grass fire raging in a hurricane of dust and smoke, there seemed an excel-

Poorman mine dumps and Fort Baker (1866 or 1867)

lent prospect for promiscuous use of powder and lead on a minute's notice." At least, that is the way the local paper, the *Owyhee Avalanche*, reported it. The editor had decided that, considering the excellent chances for a shooting war to break out, he better report the whole business conservatively.

The embattled Poorman defenders erected Fort Baker — "built of logs, with portholes and other means of defense usual in such cases." Knowing that the law almost certainly lay with them, the Hays and Ray interests were smart enough not to attack. Instead, they combined with G.C. Robbins' New York and Owyhee Mill Company to get enough financial support to bring legal action and applied in Associate Justice Milton Kelly's court for a temporary injunction. Judge Kelly arrived there October 6 to hear their complaint.

Kelly's solution to the problem of whether the Poorman was on the same vein as the Hays and Ray was simply to order additional excavation at the Poorman site to expose the vein structure. Then after hearing arguments on October 19 and 20, he granted the preliminary injunction that the Hays and Ray claimants sought against the Poorman people. Under his order, the Poorman could not be worked anymore until the matter was settled legally. Kelly's injunction was a victory for the Hays and Ray interests. Eventual development of the property confirmed that the judge was right and that the Poorman was simply part of the Hays and Ray claim. But for a generation after his decision, Milton Kelly was known to his political detractors as "bribery Kelly": The gossip was that in this litigation involving a multimillion-dollar mine, the judge most assuredly ought to have been bribed. Years later, Judge Kelly, as editor of the *Idaho Statesman*, finally got tired of the constant public harping in opposition newspapers about "bribery Kelly" and in a libel action he forced a Wood River editor to confess that those aspersions against him were entirely unfounded.

Before Judge Kelly had an opportunity to dispose of the Poorman litigation with a permanent injunction, attorneys for both sides reached a compromise that gave the greater part of the combined Hays and Ray-Poorman property to G.C. Robbins' New York and Owyhee organization, the company which had backed the Hays and Ray people financially during the suit. The Oregon Steam Navigation Company group came out with a share in the consolidation, as did the original prospectors. In the settlement, however, the New York and Owyhee listed the cost of acquiring the Poorman at $1,050,000 and its legal expense at $44,575.06. It might have been cheaper for them to have bribed the judge at that.

In addition to these costs, the claim dispute had led to much additional waste in the development of the property as a whole. A mining engineer noted that a "large amount of unnecessary work has been done on this [the Poorman] mine; one shaft sunk near the office would have been all that was necessary; but when claims are in litigation much useless work must be done to prove identity of a vein." Another result of the Poorman battle was a fairly long delay before production resumed. The attorneys agreed that the Poorman would remain closed at least until July 6, 1866, and the only active part of the property in which development continued during the interval was the Hays and Ray section of the vein.

⚒   🐴   ⚒

While the Poorman was shut down, More and Fogus continued to use several mills to handle their Oro Fino and Morning Star ore. During those months, some of the large new mills built with

This drawing of Fort Baker in 1865 was published in A.D. Richardson's *Beyond the Mississippi*.

eastern capital came into operation. The elaborate Shoenbar plant, equipped from the beginning with the efficient Washoe process, started up on November 17, 1865. The mill with its ten 660-pound stamps and a nine- to ten-inch drop, powered by a twenty-five-horsepower steam plant, could process twenty-five tons of ore in twenty-four hours. In its first two or three weeks, $20,000 in silver bricks came from this plant. Then G.C. Robbins' New York and Owyhee twenty-stamp mill, which had cost $100,000 for installation in addition to $100,000 spent in developing the mine, began to operate after a big celebration on February 7, 1866. This mill ran mainly on Hays and Ray ore. Next, on February 14, the Cosmos forty-horsepower, twenty-stamp mill joined the operating concerns. By March 9, when O.H. Purdy compiled a full report on stamp milling and quartz production to that time in Owyhee, the district had ten mills with 102 stamps. In its stamp mill capacity, Owyhee then stood second only to South Boise where bigger mills had been imported around Rocky Bar.

Large-scale mining was indeed under way by the time of Purdy's report. By then the average per ton for the mills, for which Purdy had accurate statistics, exceeded $142, an average computed on over 10,000 tons of ore from thirty-one lodes in 821 mill days, for a total value of $1,460,000. Considering that the best ore had already been shipped out, this figure was certainly a flattering one. The Gould and Curry, Virginia City's big Comstock property, was doing not much more than half that well. Of the ten mills reported, More and Fogus' 426-day operation of the Morning Star (eight-stamp mill) accounted for $1,127,617.39 of the total recovery. Although a smaller mill, the Morning Star had more than doubled the production of the others, for it had run regularly after the

Cosmos mill in 1867

fall of 1864. With the great quartz-milling boom in the spring of 1866, mining in Owyhee had become almost entirely a quartz operation. A few hydraulic giants still were making individual returns of $20,000 for the 1866 season, but most of the placer ground was being handled by Chinese miners. And there was not very much left for them.

With quartz operations going full blast in 1866, Owyhee continued to be a sober and respectable mining community. Even during the height of the Poorman explosion the fall before, Owyhee avoided the wild, boisterous atmosphere so often thought of as characteristic of mining camps. J.S. Reynolds noted:

> Ruby City is without exception the best behaved mining town in existence. Unlike other mining camps, business is closed early, and one billiard table, one hurdyhouse and two or three saloons, make all the sound to be heard after nine o'clock P.M. The reason is apparent. This is a universal laboring community, and men who do a day's work at such kinds of work as is being done here are not in the best condition at its close to keep a row going half the night.

When the large new mills started up in 1866, the three biggest ones closed operations every Sunday. Silver City was exceptional in this respect; in those days, the other stamp-milling camps did not observe Sunday closing. J.A. Chittenden, a leading Owyhee assayer, also found time to serve as Territorial Superintendent of Public Instruction and to arrange to keep the school going in Silver City. Regular church services were held, and a Sunday church school for the children was also maintained.

Except for a shutdown on April 2, 1866, to wait two or three weeks for the roads to be passable again, Owyhee stamp milling continued without interruption into the summer of 1866. Some of the mill superintendents wished that more companies had been as energetic in developing their properties as More and Fogus had been, for shortages of ore began to appear as more mills commenced operation. But More and Fogus were supplying a number of mills. Their Oro Fino property alone employed sixty or seventy men in the mine and kept forty-three stamps busy in the mills during May. By the end of June, the number of stamps in actual operation had risen to fifty-three, but this number was still less than half of the available mill capacity. Then, when the Poorman resumed mining on July 19, 1866, with a force of thirty men, it looked as if the stamp-milling companies would no longer have to wonder where their next ore would be coming from. Soon eighty-two stamps were going strong, and production was about $70,000 a week.

Less than a month after the Poorman resumed work, a major calamity set back mining in Owyhee.

Oro Fino mine

Shoenbar 10-stamp mill on Jordan Creek in 1865 or 1866

Minear mill (1864) was a mile about Silver City on Jordan Creek

More and Fogus, heretofore responsible for most of the production, failed financially on August 14. The More and Fogus disaster arose not from a failure of the Oro Fino and Morning Star properties to produce, but from the owners' financial involvement in other mines. This overcommitment to acquiring properties had paid them well to date and had been primarily responsible for what production had occurred in the Owyhee country. But now they had overextended themselves. Fogus had put a lot of money into the Rising Star at nearby Flint, which was a good property but not yet returning much. Even later in the fall, problems with milling the Rising Star ore had not been solved, and Fogus eventually sold out before successful production began there. By over-expanding, and at the same time neglecting to pay their workers and other creditors, More and Fogus found themselves broke. Attempting at first to dodge their payroll and other obligations by selling out to A.T. Minear, another mill owner, they raised a great tumult in the community when word of that deal got out on August 29. J.M. More returned to Silver City to address his multitude of creditors on September 5. He pointed out that, in his operations with Fogus,

> their business had been recklessly conducted; that more money had been expended on outside affairs and worthless men and objects to the company, than the whole indebtedness of the firm; that he did not expect to live long enough to find out all the wrongful transactions here; that of themselves the mine and mill were paying institutions; that he did not hold himself blameless for the present condition of affairs.

To the community, it looked as if "a system of dishonesty, deception and total disregard of business obligations have characterized the officers in charge" of the More and Fogus concerns. But More announced that he was placing everything he had at the disposal of those to whom he owed money. The debt amounted to $200,000. By September 8, all the creditors joined in negotiating to purchase the More and Fogus enterprises so that they could operate them to recover the debt. These arrangements, they hoped, would get the Owyhee mines and mills back into production. Even so, a host of subsidiary difficulties would remain to embarrass the community. By now the *Avalanche* noted that because both More and Fogus have "refused to pay their indebtedness, a general distrust of firms and individuals seems to be prevalent." Many workers had gone unpaid for months, and now a number of forced sales and attachments set back the district.

In an effort to bring a sound group of More and Fogus mines back into production and pay off all the labor and other obligations, the creditors, mainly workers, organized the Morning Star and Oro Fino Gold and Silver Mining Company as a cooperative venture on September 13, 1866. After

Morning Star mine and mill in July 1868

getting the Minear purchase cancelled, they bonded the property to J.M. More for him to sell in New York before July 1, 1867. Within a week, however, the new company became sufficiently distrustful of More that the deal was annulled. At this point the unpaid workers decided to assume all the other creditors' obligations and to operate the mine as a cooperative to recover their own back pay; their wage liens alone amounted to $21,891.57. Cooperative mining, though customary in placer operations, rarely had been used in quartz enterprises. In this case, the cooperative proved successful. Work, however, did not resume on the Oro Fino and Morning Star until April 28, 1867. Because these two mines had been essentially the only two Owyhee producers before the failure of More and Fogus, the area certainly was set back for a time.

Considering the disaster that had come to quartz mining in South Boise when similar failures struck the leading companies around Rocky Bar, Owyhee escaped relatively unharmed. Production actually increased decidedly for a time, because the Poorman got under way again just before the More and Fogus collapse. Yields reported by the Silver City assayers (and these figures do not account for the total Owyhee production) rose from $46,000 a month in June and July to $178,000 in August when the Poorman began to boost the total; $294,000 in September, when the Poorman accounted for most of the value and More and Fogus no longer contributed; and to over $370,000 in October. Then Owyhee production dropped off again. The Poorman, by previous arrangement, shut down again November 1, 1866, after turning out $546,691.59, processed in four mills at an expense of slightly over $90,000. Mining costs of $38,000, hauling costs of $17,000, and some other expenses still were small enough to leave a profit of $390,000 for the Poorman. An additional fourteen tons of Poorman high grade, shipped to a new plant at Newark, New Jersey, augmented the total for the second production period to about $600,000; recovery on the Newark ore ran about $4,000 per ton.

Some choice samples of the Newark ore attracted much interest when displayed in Congress that winter. Another spectacular sample shipped from New York on January 17, 1867, to the Paris International Exposition was awarded a special gold medal.

Although the Poorman had remained in actual production for less than six months in the year after its discovery, the total return for that brief time exceeded $1 million. With the rich surface concentration stripped off, the property could not continue to turn out bullion at that rate. At greater depth, the vein in its wider parts narrowed to only a few inches in width; in places where it got to be as much as five feet wide, the ore lost most of its value. Moreover, during the interval that the Poorman was shut down again after November 1, 1866, Owyhee had another setback. With the Morning Star and Oro Fino not yet able to resume production, the new Cosmos mill had a hard time surviving a series of legal attachments complicated by demonstrations of "unruly workers" who objected to not being paid for their labor.

The revival of work at the Oro Fino, conducted now by the mine workers cooperative in the interest of More and Fogus creditors, began to boost Owyhee again in April 1867. During the More and Fogus era, the Oro Fino had provided 1,200 tons averaging $33 for an additional total of $396,000. At least $60,000 more had been high-graded out of the mine, and not included in the total figure. The mine workings had been left in bad condition when More and Fogus failed. Although the workers had little in the way of resources to develop the Oro Fino, they were able to sink two shafts, discover richer ore, and pay off the liabilities (including their own back pay) by the end of the first year of their organization. After they had completed eight months of production, they had a gross of $122,000. Their ore ran from $40 to $45 a ton, and their mining and milling costs from only $25 to $27 a ton. Then, without ever having been really developed, the property was idled by more litigation and remained so for many years.

⚒ 🐴 ⚒

New life came into Owyhee in the summer of 1867 with two major finds. Charles S. Peck, who already had the Poorman discovery to his credit, traced some rich float on June 7 to a previously unsuspected vein right at Silver City. This new ledge, named the Potosi, looked as if it were a second Poorman, and efforts to develop it commenced immediately. It never did live up to its early expectations, though. Then early in

War Eagle Mountain from the east (after 1868), showing the Golden Chariot at the left, the Ida Elmore in the center, and part of the Oro Fino vein well above the buildings at the right. The Oro Fino workings are mainly beyond the border of the photograph.

September a great excitement—reminiscent of the Poorman commotion exactly two years before— swept over War Eagle Mountain, where still another spectacular vein was uncovered only a very short distance from the Oro Fino. Three different companies entered three conflicting claims, two of which were shortly consolidated. That action left D.H. Fogus and his associates with the Ida Elmore right next to Hill Beachy's syndicate with its Golden Chariot. Just as the Poorman group had previously, the Golden Chariot people argued that they were on a separate vein parallel to that of the Ida Elmore. Again, if both the Ida Elmore and the Golden Chariot properties were on the same vein, the Golden Chariot had no claim. And again, it turned out that they both were on the same vein. In the meantime, about one hundred men worked frantically, sinking shafts next to each other in order to develop separate mines as quickly as possible. Here was still another example of wasteful duplication of effort.

Just as work on these latest magnificent silver discoveries in Owyhee was getting off to a good start, the laboring force, scattered as it was among a number of different companies, decided it was about time to get together to avoid another series of mine-mismanagement calamities. At a meeting in the Owyhee County court house in Silver City, a labor union was organized on October 1, 1867.

At this time, miners' unions were quite a novelty. Outside of the Comstock where one had been established in 1863, very few existed in the western mining country. Miners from Virginia City, however, began to spread the concept of unionism to other western districts into which they migrated, and Owyhee — in many respects a second Comstock of the early days — was one of the first of these.

Because there had been no advance warning that a union was about to form, the mine operators were greatly surprised. After a brief strike that lasted four days, two basic union needs were met. The most important of these — that workers be given contracts instead of being hired informally — was intended to help protect Owyhee mine labor from being left unpaid while mine proceeds went into other projects. Another union request — that wages be paid in bullion rather than in depreciated greenbacks — also seems to have been honored. Wage increases, however, were resisted somewhat more successfully. The $5-a-day rate at Silver City was regarded as rather high already and had been maintained so far because of the remoteness of the district. California wages, by contract, ranged from $2 to $3.50 a day. After considering a demand for $6, the union decided to press for $5.50, a rate which they did not obtain throughout the district. Some companies agreed to the increases, but others

did not yield at all. Without any tradition of long strikes or other union activity, Owyhee miners decided to accept what increases they had gotten in these few days and returned to work. Their union, however, had been organized to contribute to miners' welfare in many other ways than in obtaining higher wages. The union movement remained important among the Owyhee miners during the next decade.

⚒   🐴   ⚒

Once the strike was over, work on the Golden Chariot and the Ida Elmore resumed with renewed energy. Rather than risk heavy expenses and losses through protracted litigation, the rival interests of Hill Beachy and D.H. Fogus agreed to a division of the new discovery. They arranged to leave a neutral ground between their two operations so that they would not be running into each other. This deal worked satisfactory until late in February 1868, when the Golden Chariot violated the neutral ground and broke through into

Idaho Hotel and Wells Fargo office, Silver City

the Ida Elmore tunnels. At this stage, both sides armed for conflict. Actual hostilities did not amount to much for a month, although considerable tension and sporadic shooting kept everyone nervous. Then on March 25, 1868, the Golden Chariot forces advanced to the Ida Elmore shaft in an offensive marked by heavy firing that threatened to shatter and break up the timbering and bury the underground belligerents. Unable to retain control underground, D.H. Fogus sent out for reinforcements with which he hoped to gain the upper hand on the surface. Quite a collection of toughs, some even from Nevada, answered his call, and the Owyhee War (as this battle was known) assumed serious proportions.

Only two casualties attended the underground advance of the Golden Chariot, surprisingly light losses considering the heavy firing in the tunnels. When the Golden Chariot group had gained control of D.H. Fogus' Ida Elmore shaft, and held onto it, nothing much more in the way of fighting was possible there. But should Fogus succeed in transferring the battlefield to the surface, where he might easily besiege the Golden Chariot, the whole war might take on quite a different aspect. No one in Owyhee seemed to have a good answer for what to do. Some criticized the county sheriff for not going in and stopping the whole affair, although none of them seemed ready to volunteer their help. Just how the sheriff, all by himself, was going to stop a hundred armed men from firing on each other in underground workings was never proposed. The sheriff did have the presence of mind to shut down all the local saloons the night hostilities broke out. Beyond that, he scarcely knew what to do. The fighting so far had been going on inside the mines, where each party claimed that it was shooting to protect its own private property from intruders and burglars. In fact, each side felt that both mines in their entirety were its own private property, once the truce had ended. By now, the problems in having two different companies, each thinking they owned the same mine, had become insuperable.

When the first hasty reports of the Owyhee War reached Boise, Governor D.W. Ballard concluded that firm action was needed. Two casualties had resulted from the initial skirmish, and law and order seemed to him to have broken down. After dispatching Idaho's most renowned deputy marshal and Indian fighter, Orlando Robbins, to the battleground with a proclamation commanding both sides to desist and to settle the dispute according to the processes of law, Ballard himself set out for the scene of hostilities. In a record six-hour trip Robbins reached Silver City, consulted the sheriff, rounded

up the leaders of the two companies, and within an hour of his arrival on March 26, read them the proclamation. No one in Owyhee had asked Governor Ballard to intervene, but the results of this effort were certainly effective.

By late that night, a new agreement had been reached, with formal deeds drawn, so that the matter did not even have to go to court. Unfortunately, during a drunken brawl on April 1, J. Marion More became the final casualty of the war. More's friends, in turn, were about to lynch their Golden Chariot opponents, but Governor Ballard, addressing the citizens of Silver City on April 2, insisted that the law continue to take its course. Matters looked so threatening that the Governor at this point summoned troops from Fort Boise. Marching to Owyhee with a brass cannon, ninety-five soldiers occupied Silver City from April 4 to 8. But, by then, largely as a result of Ballard's firm action, the Owyhee War was over.

If the deals which led to the Owyhee War were intended to avoid the expenses of mining litigation, it is dubious just how much they saved. For the Ida Elmore, $200,000 out of the $600,000 of the first year's production is reported to have gone into paying the cost of the war and the litigation. The Golden Chariot, which in its first year realized only $200,000 because it had shut down to install new hoisting works, had to devote all this initial gain to covering the expenses of the battle. The conflict served to emphasize again the waste and inefficiency of having rival companies develop short, adjacent segments of the same vein with separate and duplicating shafts, hoisting works, and company organizations.

※   🐎   ※

By the time the Owyhee War was over and normal operations had resumed. Production went back up to more than $200,000 a month, to the level it had been since the new discoveries began to affect the figures significantly. Concurrently, an important improvement in technology came to Silver City in the summer of 1868 in the use of dynamite to reduce expense and increase production. Shipments of gold and silver from Silver City totaled a little over $3 million by July 8, 1868, and the following year that amount increased another million. These figures represent quartz production primarily and show that in spite of a number of failures, the constant appearance of bonanzas kept the better Owyhee stamp mills going most of the time. By 1869 three companies had grossed over $1 million each: More and Fogus (using both the Oro Fino and the Morning Star), the Poorman, and the Ida Elmore.

Poorman mine development included a series of openings along an extremely rich silver vein in about 1867.

Silver City and War Eagle Mountain looked like this in 1868: the Morning Star mill shows on the hill to the left, and the old Owyhee County courthouse, which was built in 1867 and burned in 1884, is by itself above the town left of center. The Lincoln mill is in the right center foreground.

Washington Street, Silver City

Silver City parade (after 1890)

Two other large operations, the Ainsworth and the Golden Chariot, had done well over $500,000. And although there had been quite a group of smaller producers, these five accounted for by far the greatest part of the $4.8 or $4.9 million of bullion shipments from Owyhee by the end of 1869 (this total does not include More's shipments which are represented in the company production totals). Of the others, the Rising Star at Flint looked promising. Purchased by George Hearst and his San Francisco associates in December 1867, that property had serious recovery problems in the process of treating ore which assayed very well. Smelting was required at Flint, and the technology had not yet been developed adequately.

Silver City prospered sufficiently by 1869 to maintain two newspapers for a time. Published originally in Ruby City before the Poorman War, the weekly *Owyhee Avalanche* had moved a short distance to Silver City. Then the Owyhee *Semi-weekly Tidal Wave* offered competition to the *Avalanche* for a time, although the two eventually had to be consolidated into a weekly *Avalanche and Tidal Wave*. By that time, Idaho had only four newspapers left, with the others in Boise, Idaho City, and Lewiston. A succession of influential editors and publishers maintained the importance of the surviving *Avalanche* for many years.

In relation to future production, the Owyhee mines had to wait for improved technology and transportation, particularly railroad service which came much later. Even the relatively early producers, for the most part, had not begun to realize their expected worth. The series of easily mined surface bonanzas had kept stamp milling from failing in Owyhee in the same way that it had failed in South Boise during those same years. But the lack of adequate investment capital had retarded both the development of the mines and in some operations the construction of efficient mills. Although several of the big producers were rich enough in the beginning to pay their own way, including their development costs, the results of financing by

War Eagle Hotel, Silver City

having mine proceeds pay for development was an improper procedure that in a number of mines created problems not overcome for years, if ever. Only about one-eighth of the total production of Owyhee quartz may be attributed to the early boom years through 1869. During that time, work had been limited almost entirely to high-grade veins near the surface on War Eagle Mountain. Until the great bulk of lower grade ore, mainly in Florida Mountain, could be processed economically, Owyhee did not realize anything like its full potential.

Work on high-grade War Eagle properties continued until the failure of the Bank of California brought a general mining collapse at Silver City by 1876. Financial problems arising from the Panic of 1873 did not disturb Owyhee too seriously in advance of a San Francisco banking fiasco two years later. On the contrary, the arrival of telegraph service in 1874 allowed Silver City's newspaper to become the *Owyhee Daily Avalanche* for a time. A rush to a great new lode discovery about six miles west down Jordan Creek near Wagontown, inspired in part by a flattering $9,425.34-a-ton assay of a high-grade sample, led to even greater optimism in May 1875. This important property, later developed by Joseph R. DeLamar, eventually became Idaho's major silver mine outside of the Coeur d'Alene region. Even after the Bank of California failed on August 26, the Oro Fino, the Golden Chariot, and the Poorman continued to produce. Finally in the summer of 1876, Silver City's labor union (which had shown considerable strength in driving out Chinese miners in 1873) had to take a firm stand against the company's failure to pay miners because of cash flow problems during that financially troubled era. Miners at the Golden Chariot held their superintendent hostage after they had gone unpaid too long. Violent battles over ownership of the Empire and the Illinois Central brought additional hardship in 1877. Soon most of Owyhee's mines had to close until superior management, more funds, greater technology, and better transportation enabled the district to exploit the large, low-grade Florida Mountain and DeLamar properties. This transition from high-grade War Eagle lodes extended over most of a decade.

Although W.H. Dewey's operations at the Black Jack in Florida Mountain forecast a new direction for Owyhee development by the beginning of 1878, all major mines there were shut down again in the summer of 1880. New investors came in with purchases of major properties such as the Morning Star in August 1882 and with the installation of speculative mills at Wagontown that fall. More capital was required, however. Finally J.R. DeLamar bought John A. Wilson's interest in an inadequately funded Wagontown property for $30,500 in September 1886.

⚒ 🐴 ⚒

DeLamar, a Colorado developer who had investigated mining engineering sufficiently to learn how to manage a property competently, explored Wilson's lode at Wagontown and introduced improved milling processes that transformed

W. H. Dewey's Black Jack mine

Dewey Hotel and mill (c. 1900)

W. H. Dewey's freight team

DeLamar Hotel in 1890

DeLamar miners had their own concert band in 1900

DeLamar mill (c. 1900)

mining there. He acquired a Silver City mill to handle his DeLamar ore until 1888, when he arranged to enlarge his mill which he removed to DeLamar the next spring. In January 1890, four months after milling resumed, he purchased the remainder of his mine for $500,000 prior to selling his entire property the next year to British capitalists for $463,000 cash along with 105,000 shares of stock in the new London corporation. Unlike any other British mining investment in Idaho, DeLamar's property returned a profit. Dividends amounting to 267 percent on this property alone covered all British investment losses in Idaho mines.

While the DeLamar mine was active, W.H. Dewey continued to develop the Black Jack. In November 1892 the Trade Dollar mill opened at the Silver City end of Florida Mountain. Flint was also productive during this era, and additional British capital helped maintain other Silver City properties as well. (Following DeLamar's sale, another $600,000 Silver City transaction had helped out in the summer of 1891.) Even the Poorman reopened for six months in 1894, until the loss of the Poorman mill in a fire halted operations. Altogether, British investors ran the DeLamar mill for fourteen years. Although Florida Mountain mining was suspended in 1897 for a time, production resumed after a merger of the Trade Dollar and Black Jack operations in March 1899. This combination led to the development of an interconnected five-mile Florida Mountain tunnel system which ran all the way from Silver City to Dewey. After Trade Dollar Consolidated installed an electric power plant at Swan Falls in 1901, Silver City continued as a progressive mining camp with major production until 1912. With a population of 976 at Silver City and 876 at DeLamar in 1900, compared with 583 and 438 in 1890, Owyhee's mines reached new heights unmatched in earlier years. Even though large-scale, low-grade production after 1886 did not generate the wild excitement of the Poorman

and Golden Chariot era, more gold and silver came out of later operations.

Limited mining in Owyhee continued in the Depression after 1929. After a 1934 referendum that removed county government offices to Murphy, Silver City did not have much of a future as a permanent settlement. DeLamar went into a still greater decline until arrangements were completed in 1976 to reopen that lode as an open-pit operation at a cost of $22 million for development and recovery facilities. In only 2½ years (out of a planned fifteen-year production period) following resumption of mining in April 1977, DeLamar turned out 19,000 ounces of gold and 2 million ounces of silver. Gold prices close to $600 an ounce and silver values at $16 an ounce made mining there profitable again. Even when gold declined and silver fell to half that level, DeLamar's production increased to 2,060 tons daily in 1980, with 1,660,000 ounces of silver and 18,000 ounces of gold that year. Dome mines purchased that property for $28 million in 1980, with ore sufficient to produce 2,500,000 ounces of silver annually for twenty years. With large-scale twentieth century production, Owyhee is scheduled to yield a great deal more than its already important historic development.

DEADWOOD

Excitement over placers in Deadwood Basin attracted several parties in the early summer of 1863, and the *Golden Age* of July 15 announced that prospects ranged as high as 12½ cents to the pan. A company of Frenchmen worked most of the summer of 1864, and several placer finds on the South Fork of the Payette River, not far from the

A redesigned DeLamar Hotel served miners after 1900

A DeLamar miner's house in 1910 was throughly modern

mouth of the Deadwood, kept prospectors in the vicinity. Interest in the Deadwood placers grew that fall when a mining district was organized on October 17. Reports of finds running 50 cents to $1 per pan brought many men there from Boise Basin. One company made $600 hundred dollars in one day, but the commotion was a temporary one. When mining at Deadwood began again in 1867, the district was regarded as brand new.

Nathan Smith, one of Idaho's most distinguished prospectors and a member of the Florence discovery party, revived the Deadwood mines. The miners' meeting over which he presided on August 16, 1867, and in which a new district was organized, may be regarded as the serious beginning of Deadwood placer operations. When the story of the new discoveries reached Idaho City, a stampede to Deadwood resulted on September 8-9, 1867. J. Marion More's party found a gulch that prospected 50 cents to $2 to the pan, and two or three other gulches also promised to yield well. After James A. Pinney, postmaster in Idaho City, returned on September 14 with a good report, it was hoped for a time that Deadwood would build up Idaho placer

mining to something like that of the earlier boom days. By the time mining ended for the fall, however, only four gulches had proved workable. Deadwood City already was "quite a little town." But aside from arranging to construct a substantial ditch for use in the dry gulches the next spring, little could be done that fall.

Mining commenced in earnest about the beginning of May 1868. One hundred men went to work in the older (August 1867) placers, and about thirty prepared to open up a new placer area as soon as the snow melted. One three-man company in the old snow-free district cleaned up $5,000 with a giant hydraulic in two weeks, and production increased when the new 300- to 400-inch ditch came into use on May 4 (water was supplied at what was regarded as a reasonable rate of 50 to 65 cents an inch for twenty-four hours). At the very beginning of the season, two men in the newer placers made $212 in two days, and in the middle of May two men took out $70 in one day with a rocker in a dry gulch. The average daily production for the most successful company reached $100 during the season that ended early in July. Those were the high

Silver City, September 24, 1895

Camp at Deadwood, August 1924

Hall-Interstate mill and Bunker Hill and Sullivan office at Deadwood basin (1924)

figures for the area, but even so, placers were worked easily and promised to last for more than one season.

Quartz possibilities in Deadwood Basin pointed the way to the main future production of the district. On May 16, 1868, one prospector worked a rich outcrop at the rate of sixty-two ounces to the ton, using a "rough process," and another good vein was discovered on June 2. Then in July, J.G. Bohlen, whose experience in the mining country had been limited to running the Idaho City Dancing School, astonished everyone with still another very rich quartz ledge. These quartz prospects could not come into production instantly, but they finally accounted for most of the Deadwood yield.

Interest in Deadwood diminished in 1869 with the gold rush to Loon Creek, and by 1876 Deadwood City had become a ghost town. Mining finally resumed there from 1924 to 1932. By 1947, a Deadwood lead-zinc property had yielded about $1 million.

LONG VALLEY

Long Valley placer discoveries would have commanded more attention in 1863, if Boise Basin had not offered so great an attraction for prospectors. Miners stampeding from Lewiston along Packer John's Boise trail noticed placers at Copeland (located presumably on Gold Fork and named for their finder) as they went through Long Valley in 1863. By June of 1864, Gold Fork's only mining company had recovered $1,600 in fine gold from a hill claim that produced from $16 to $20 a day that spring. Other miners were out prospecting there, but high water interfered with work on stream placers. Two other early districts also gained attention. A camp at Lake City (presumably near

Payette Lake) had created interest sufficient to induce miners to import a sawmill. But long before 1870, Lake City had become a ghost town of fallen-down cabins, although an abandoned saw mill was still in good shape in 1872. Between Copeland and Lake City, another camp known as Hawkeye emerged about ten miles from Copeland. Across the ridge beyond Copeland, scattered miners also placered along the South Fork of the Salmon River.

When other gold rushes ceased to divert interest from Long Valley, Copeland did a little better. Deep placers and the lack of water retarded development, so almost a decade went by before Gold Fork placers could be largely worked out. Copeland's bar, about a hundred feet wide, had to be dug to a depth of forty feet. The lack of water, except during a short spring season, held back development. By 1870 a modest mining community was engaged in constructing ditches and whipsawing lumber. Sixteen men and one woman spent the next winter in that isolated camp, continuing their preparations. By 1872 they had recovered most of the gold they were going to be able to save. By that time, interest had shifted to Hawkeye where new ditches were dug. Enough gulches near Gold Fork paid $6 to $8 a day so that a limited amount of work continued for many years. By 1879, John Kennaly of Boise had completed a nine-mile ditch which paid off, and he wanted to extend it to twenty miles to reach more ground.

Occasional excitements continued to follow new discoveries. A stampede to Long Valley in the summer of 1886 stirred up additional interest, but the excitement at Yellow Pine that same summer lacked foundation and discouraged those who wanted to expand mining on the South Fork of the Salmon River. Eventually in 1892 the Standard Oil Company of Ohio acquired 240 acres on or near Gold Fork for placer development.

A different kind of dredging production came to

Long Valley several decades later. When the principal imports of monazite (a source of rare earth elements used in petroleum catalysts, ceramics and glass additives, and electronics) were cut off in 1946 and 1950, an investigation of localities in Idaho led to extensive Long Valley dredging for monazite beginning in January 1951. Three dredges operated that year. Finally one capsized and sank in 1953. The lack of a market led to suspension of both remaining operations in 1955. By that time, monazite production had reached about $2 million. By-products raised total production value to $3.5 million before mining ceased.

STANLEY BASIN AND ROBINSON BAR

So many prospecting parties set out from Idaho City, Placerville, Rocky Bar, and other established southern Idaho camps that many regional streams and ridges had been examined for gold and possibly silver within a year or two after 1862. Reports of fabulous new discoveries circulated through established mining centers throughout the West. Occasionally such rumors had foundation, although successful prospectors often went to great effort to conceal any actual discovery in order to avoid the plague of claim jumpers who could be expected to invade a promising new district. More often than not, reports of valuable new mines proved erroneous. More than a few premature allegations of mineral discovery related to places where mining was eventually to occur. Other mines with imaginary antecedents have failed to materialize in more than a century. Mineral possibilities around Hailey, for example, attracted notice almost two decades before serious development, whereas Stanley and Little Smoky had modest beginnings that traced back to 1863 and 1864, even though most actual work there came after Bellevue and Hailey attained prominence. Little reliable evidence may be adduced to clarify occasional later recollections concerning early prospecting of the Sawtooth country. A 1904 report of C.E. Jones and an account of Frank R. Coffin (a member of John Stanley's discovery party) provide essential information concerning otherwise obscure mineral exploration of the Sawtooth wilderness.

John Stanley set out with a group of sturdy prospectors from Warren after an interesting July 4 celebration in 1863 to investigate Bear Valley and Stanley Basin. Modest gold prospects around Stanley (named for the expedition's captain) failed to interest anyone then. Most of Stanley's crew chose to go back, but he took a small group on to discover Atlanta on their way to Idaho City. The next spring, about seventy men set out from Boise to follow up the Stanley Basin discovery. They created a great Wood River excitement, described in some detail in the *Boise News* of April 30, 1864:

> A party of some seventy men left here about three weeks ago on a prospecting excitement, with such mysterious and undefined purposes, that we have been thus far unable to ascertain their destination. It seems that late last Fall, a party of fifteen prospectors, who had been out about the head waters of Wood river and probably Salmon, came in after provisions, and loading their animals, attempted to get back to their new mines with their freight, but were unable to get further than the vicinity of Camas Prairie with their animals. They then made themselves snowshoes, sleds, &c., and taking what provisions they could, went on. Parties followed them to this point, where their trail was lost. As the spring advanced, and the prospect of getting over the mountains grew more favorable, the present excitement commenced.
>
> The party proceeded to within a short distance of Big Camas Prairie, and finding that they were too early to get across the mountains, a portion of the company concluded to return. While in camp at this place, an Indian came to them and told them there were a great many Indians within ten or twelve miles on the opposite side of the prairie, and desired to be friendly. Afterwards a party of four Indians came into camp, carrying a flag of truce. They could not talk English, and were unable to let their business be known. A portion of the company, some five or six, not being satisfied of their good intentions, pursued them and shot one of the Indians. In returning from the chase, a gun, in the hands of one of the parties, was accidentally discharged, striking Jesse Peters, making a severe if not mortal wound in his hip.
>
> The party then scattered, some returning, and some going on with the hope of getting through. The new diggings are supposed to be in a basin near the head of Wood and Salmon rivers.

Coming up Goodale's cutoff to Little Camas Prairie, this group ascended the South Fork of the Boise past the South Boise mines and Little Smoky. Here they met with some success, although the Little Smoky placers did not detain them too long. They worked their way along an old Indian trail through deep snow over the divide to the upper Salmon. Continuing north through Stanley Basin, they camped near the future site of the town of Stanley, nearly four miles from their long-sought mines. To their consternation, the camp grew to around two hundred miners that evening. Exaggerated accounts of the richness of these prospects had begun to circulate, and too many gold hunters were getting interested. The next morning, a grand stampede of eager prospectors galloped over to the new placers. The original party decided they would have to locate claims before any more hopefuls showed up. Upon arriving at the discovery site, they found that enough miners had gone ahead to locate all the ground. Nothing remained for any of the stampeders. Some went on to find galena outcrops on Wood River and copper prospects on Lost River. But these possibilities did not interest anyone in 1864. Later miners turned out $15 million in the Lost River copper properties around Mackay, and Wood River galena eventually proved to be worth over $60 million.

Following the 1866 Leesburg gold rush, interest grew in Stanley Basin. River bars for a hundred miles below Stanley also proved attractive. In 1867, mining finally got under way around Stanley, and Robinson Bar gained local fame. Giant hydraulic operations had commenced in 1868 along a six-mile segment of the Robinson Bar placer. Large boulders and rocky ground, however, greatly limited that effort. Like other upper Salmon River bars, those placers did not respond to conventional mining methods. Tunnels were employed so that gravel could be dug out. Work could go on in winter as well as in summer, allowing Robinson Bar to retain a considerable mining camp for a decade, until more exciting Yankee Fork properties diverted most Salmon River bar miners to Bonanza in 1878. Until then, Stanley mining district—with upper Marsh and Valley creek placers active in addition to workings at Robinson Bar—underwent modest development. Stanley district mines offered employment to twenty-nine miners on June 1, 1870, in spite of a variety of problems which discouraged activity there. Valley Creek and Marsh Creek between Stanley and Cape Horn provided "quite an extensive area of mining ground. . .which can only be slowly worked, on account of the scarcity of water." In 1871, two miners worked "their claims by the hydraulic method, for about two months in the spring, and they take out about $20 per day to the hand. For the balance of the year they take it easy, having plenty of time to hunt and fish, and having unequaled hunting and fishing ground."

Two years later, A.P. Challis, who had come through Stanley Basin almost a decade earlier as a member of John Stanley's discovery party, and Henry Sturkey began to develop their placers at Stanley. Seven years' work went into getting the Stanley placers ready for production. In 1874, miners were busy for a hundred miles along the Salmon River, mainly above Bay Horse past Robinson Bar and Stanley. A short placer season (limited by the lack of water most of the year) kept

Challis and his associates at work around Stanley until the end of the century. Finally a dredge got started on September 24, 1899, and their modest, yet dependable, operations were enlarged to a production of $6,400 annually. Even at that, only twenty-five men could work profitably in the area. Lode discoveries in 1897 and 1902 led to the importation of a stamp mill, which was auctioned at a sheriff's sale late in 1904. After that fiasco, and after forty years of preparation, mining in Stanley Basin finally became more prosperous.

ATLANTA

Of the larger mining areas in Idaho, Atlanta got off to one of the least conspicuous starts. Several quiet years passed before preparation for big production could be made in Atlanta, and then decades of false starts and failures of one sort or another followed the initial setbacks that characterized stamp milling in most Idaho mining camps. Really successful large-scale mining of the Atlanta lode did not come until 1932. A large quantity of good ore had been processed before that time, but not with the same relatively acceptable results that had characterized similar mining in Rocky Bar, Owyhee, and other such districts after 1886.

A remnant of John Stanley's prospecting party discovered gold near Atlanta after panning some inconsequential finds on a trip through Bear Valley, Cape Horn, and Stanley Basin before heading over the Sawtooth Range into the Middle Fork of the Boise. Arriving in Idaho City early in August 1863, Stanley's party kept all its discoveries secret. Or at least the members tried to. Prospectors who wanted to conceal a substantial gold discovery often attracted attention anyway. Shortly after August 8, a great rush of miners set out to Atlanta in search of Stanley's new placers. An Idaho City correspondent reported on August 12 to the *Washington Statesman* (Walla Walla) that

> we have had quite an excitement about the new diggings on the Middle Fork of Boise, about 100 miles from here. Hundreds of miners have left on the report that the mines yeild [sic] from 10 to 50c. to the pan.
>
> The merchants here are reaping a rich harvest on their goods "over the left." Shovels retail at $2.20; picks with handles $2.50; beans 25c; bacon 35 to 40; Sugar 30 to 40; E.B. Syrup, 5 gallon kegs, 15 to $16, per keg, coffee 50c. Liquors are in lar[g]e supply and selling at 3.50 to $5 per gallon; long legged rubber boots are retailing at $10 per pair. The only article in demand just now is flour: it however, only 3 weeks ago was selling for $20 per hundred, and

within the past week it has been sold at $40 per hundred. This sudden rise will undoubtedly cause the shipment of large quantities not only from the Columbia River, but parties have gone to Salt Lake to bring in forty or fifty tons. Some of my friends have left for the new mines, and by the next express I shall be able to give you definite information from there.

Within a week, a host of dejected gold seekers returned to Idaho City to tell of failing to find mineral wealth around Atlanta.

> The stampede of miners of the Middle Fork of Boise, upon news of new and rich discoveries there a short time ago, has resulted in stampeding back again — false alarm. Proceeds of the trip: torn shirts and ragged breeches. — Most of the steamboated miners who returned, look too wolfish to interrogate as to particulars of the journey. The reported rich diggings were about 65 to 80 miles from Bannock. The very air is said to have been blue with their curses, on the return trip.

A few good Middle Fork placer bars were noted later in August, but Stanley's party thus succeeded in keeping its Atlanta discovery secret after all. But when they got equipped to return with a much larger party in 1864, their preparations caused a great deal of curiosity. More by accident than design, however, their Atlanta find remained entirely confidential. Unimportant finds around Stanley Basin obscured significant prospects around Atlanta. When a substantial party set out for Stanley in April, 1864, a large group of miners tagged along to participate in the hunt for new gold fields. That left Atlanta's explorers free to search out Idaho's major gold lode in secrecy. News of mining prospects around Atlanta did not get out for most of a year. Finally, after a Yuba River mining district was organized on July 20, 1864, Atlanta began to attract little notice.

Placer mining of consequence got under way in the Atlanta region very late in September or early in October 1864. About one hundred men were working by mid-October, reported Charles H. Rogers of Happy Camp, a nearby South Boise community. Yuba River claims were fairly good, and it was thought the river would pay $10 a day over its entire fifteen-mile length. The highest rocker paid $30 a day, but the average, naturally, was much less. There were some complications, though. Except for two bars, all the Yuba River placers were in a deep canyon that was hard to work. Aside from some Frenchmen who had done quite well, those who hoped to make their fortunes on Yuba River were having trouble. Most of the stream could not even be prospected adequately because of the swift current. It was clear to all that

the Yuba River was not a poor man's stream. Only with considerable effort and expense could mining be developed there.

An astonishing quartz discovery in November changed the prospects of the Yuba River mines. Called at first the Eagle of the Light after the Nez Perce leader known primarily for his utter opposition to white miners in northern Idaho, this discovery soon became known as the Atlanta lode. This name, selected by Confederate refugees who predominated in the early Yuba River mines, was chosen in honor of Confederate General John B. Hood's reputed great military victory over General W.T. Sherman in the battle of Atlanta that summer. Hood's self-serving public relations, in this case, had greatly exceeded his military success, and some time was to pass before the correct news of Sherman's triumph over Hood and his march to the sea finally penetrated the mountain fastness of Yuba River. Even so, the name remained the Atlanta lode, and eventually the major camp there was known as Atlanta also.

From the very beginning, the Atlanta lode looked big: "It is truly a mammoth ledge, and in one place crops out about 60 feet above the surface, and is at least a quarter mile wide, while thousands and probably millions of tons of quartz have fallen from its rugged sides and scattered over the mountain." A great rush to Atlanta on November 19, 1864, of some twenty prospectors resulted in a rash of additional discoveries. Claims 200 feet long were taken up all over the lode, although it was too late in the season to do much more than locate claims.

Preparing mainly for the next Yuba River placer season, about forty men spent the extremely hard and long winter of 1864 and 1865 on Yuba River. The new mining season, therefore, was late in starting. By June 1, C.W. Walker reported from Bedrock City — a metropolis, apparently, of one cabin — that most of the mines were still not opened. But the Cavanaugh claim which had been worked the previous fall was going as well as it had before. Extensive ditches were under construction, the most ambitious of which was Mattingley's; this ditch from the mouth of Quartz Gulch (later the site of Atlanta) around to Yuba River gained an elevation of 109 feet, 11 inches. Mattingley was building a large frame house where his ditch began, and that area was regarded as the best townsite available. Two earlier towns — Alturas City at the mouth of Yuba River, and Yuba City, upstream at the mouth of Grouse Creek near the Atlanta lode — already had been established, but neither amounted to much, though Alturas City boasted a store.

Yuba River placers proved disappointing in 1865. But about two hundred men were able to work on the Middle Fork near the mouth of the Yuba River during the season. The big attraction, however, was plainly the Atlanta lode. Traced for 1½ miles of length, it ranged from fifteen to thirty feet wide, or greater, depending upon the optimism of the prospector. William J. Libby decided that summer that Atlanta really needed a mill, and in anticipation of bringing one in, he began to work on a road from Rocky Bar. His mill, while nevertheless on its way across the plains, could not be expected to reach Yuba River until the road might be completed the next summer. Then in November, W.R. DeFrees discovered the Greenback about two miles from the Atlanta lode. Convinced that it had great worth, he rode out to California to obtain a sawmill and a stamp mill. Atlanta, therefore, was to be equipped with two mills in 1866.

⚒ 🐎 ⚒

Additional lode discoveries in the summer of 1866 brought increasing excitement to Atlanta. The Leonora in July looked the best. It did not resemble the Atlanta lode at all. It was in gold rather than silver, and although it ranged from very thin up to two inches in width, in its thinner parts it was almost a solid sheet of gold; indeed, the owners thought that they took out almost $10,000 worth of ore in a few hours one day. A $230 nugget, found in Quartz Gulch between the Leonora and the Atlanta lode, also stirred up some interest then.

Mills to test the mines were also reaching the country. M.C. Brown, who had taken over W.J. Libby's pioneer stamp-mill project, got the original mill ready for operation in September 1866. Then W.R. DeFrees, whose stamp mill had left Chico on June 9, had his sawmill running and his stamp mill nearly completed in October. Another 1866 enterprise, that of J.W. O'Neal, started on a refreshingly different principle. He chose to develop his mine first, and then to bring in a mill if the ore to justify one proved to be available. O'Neal, who had learned from the Rocky Bar disasters, pointed out that if his enterprise, supported by capital from Reading, Pennsylvania, should fail, the only loss would be the expense of prospecting.

By the end of the 1866 season, Atlanta had developed to the point that serious quartz production might presumably start. The major deficiency in the arrangement then was that the ore was primarily silver, whereas the stamp mills brought in were equipped only to recover gold. The Washoe process, developed on the Comstock lode and already in use in Silver City, was suitable for Atlanta also, but it was not tried there until after 1869.

Extensive development by the current standards preceded milling experiments at Atlanta. Lessons learned across Bald Mountain Ridge at Rocky Bar in 1866 were put to use in Atlanta in 1867. Not just two or three but quite a number of companies ran tunnels and shafts and got something of an idea of what kind of ore they might expect before trying to mill it.

Their work led to a number of gold excitements. Nelson Davis found excellent examples in a new property near the Minerva late in 1866, only one of a number of major discoveries that fall which promised to do well in 1867. W.R. DeFrees found his part of the Atlanta lode to have a width of twenty-two feet and a depth of ninety feet, with an average value of $75 a ton. After Matthew Graham's company had "spent a large amount of money" driving tunnels and shafts in the Lucy Phillips, they concluded that the property justified installing a twenty-stamp mill. Similarly William Clemens, who proved the Minerva to be rich, planned to bring in a mill in the spring of 1867. J.W. O'Neal, having thoroughly prospected his ledge, came to the same conclusion. All of these plans, formulated by the end of February 1867, promised an eventful future. The Atlanta lode, particularly, had great value. Two hundred feet in the fourth extension sold for $10,000 in gold coin, and the total development by February amounted to four thousand feet in length. Not only did some of the assays from the Atlanta lode, ranging as high as $11,000 a ton, surpass those of the celebrated Poorman and Owyhee, but the lode itself was also incomparably larger.

Unlike Rocky Bar and Silver City, Atlanta resembled the Comstock in having a large lode rather than in consisting of a system of fissure veins such as those of Silver City. The Atlanta lode, moreover, was suitable for deep mining in the tradition established at the Comstock and tried unsuccessfully through the latter nineteenth century in numerous districts such as Rocky Bar, where the ores simply did not persist to great depth.

Much of the error in estimating the potential of mining areas in these years came from assuming that rich values would extend to great depth, as they did at the Comstock. Moreover, after the Great Bonanza was struck at a depth of almost 1,200 feet in the Consolidated Virginia on the Comstock in 1873, miners all over the West anticipated similar finds. Atlanta did not have anything of the combined magnitude and value to match the Great Bonanza on the Comstock, as yet undiscovered in 1867, but Atlanta did have a lode containing an enormous quantity of ore amenable to the Washoe recovery process of the Comstock.

On that account, large stamp mills could be employed there to greater advantage than in many other districts.

Atlanta suffered, however, by being even more remote than such isolated districts as Rocky Bar. A high ridge separated the two camps, and a long-projected water-grade road from Boise up the Middle Fork to Atlanta did not materialize until 1938. Stamp milling at Rocky Bar had suffered severe reverses by 1867, in part from the difficulty of transportation. Atlanta was decidedly worse off in that important respect. Ironically, this early southern Idaho camp, which had the great quantity of ore that might have been developed to supply a really large stamp-mill capacity, was located so unfavorably that mills could be installed and operated only with great difficulty. Places such as Rocky Bar, not quite as hard to get to, had large stamp mills but lacked the potential ore reserves of the Atlanta lode.

⚒ 🐎 ⚒

Actual stamp milling finally commenced in Atlanta in July 1867. At last the new lodes no longer had to depend upon arastras which were already in operation. W.R. DeFrees' mill, the one which started in July, did not show to great advantage at first. Indeed, it broke down, and although it was expected to resume on July 6, more than a month went by while the operators worked to get it going again. Then a preliminary fifty-six-pound lot was managed by mid-August, with a promising return of thirty-two ounces. At last, an initial twenty-ton run commenced on August 19 so successfully that the mill had to stop every two hours to clear the amalgamator until the entire supply of mercury (eight flasks) was exhausted. (The amalgamating practice at that time apparently did not include retorting to recover mercury at the mill, although Silver City mills retorted their amalgam.) Because the better ore was not yet being milled, this beginning looked promising indeed. After the first 9½ tons—the part actually worked before the mercury supply ran out—produced about $6,000, W.R. DeFrees left Boise at the end of August on a trip east to raise capital to develop what he now could represent as a proven mine and mill. His milling process, however, was probably losing three-fourths of the value, including all of the silver. But even at that, it would pay.

A second run of sixteen tons, begun after more mercury had been obtained from Idaho City, was completed on September 17. Although this run was thought to be much better than the first, work in earnest waited for another year while capital was being raised. The district was not shut down entirely, however; the mill continued to test batches of ore for two other companies, while an arastra continued to grind out twelve tons of gold ore worth about $700 that fall. An energetic miner, William Clemens, pounded up a quarter of a ton of Minerva ore in a hand mortar and recovered $250 while sampling. Thus, Atlanta had gotten off to a reasonably encouraging start, although a mining engineer writing for J. Ross Browne's *Report. . .on the Mineral Resources of States and Territories of the Rocky Mountains* felt that at the end of the 1867 season "the depredations of Indians in neighboring districts, the mismanagement, the want of skill and proper knowledge of the business, and the incompetency of agents and superintendents with the misapplication of capital, had done much to retard the development of the Atlanta mines." It was conceded, though, that DeFrees' water-power mill was working with "very satisfactory" results, "although imperfectly, from want of proper appliances and skill."

In the fall of 1867, when things generally looked good in Atlanta, an unexpected calamity befell J.W. O'Neal's twenty-stamp mill. The Alturas Mining Company, for which O'Neal had found a mill site and prepared to erect a building, had a twenty-ton mill well on the way to the lode. But the president of the company, an easterner with no background in mining whatsoever, reached Rocky Bar by the time that the mill had arrived at Junction Bar (modern Featherville), eight miles below. Although on his way to Yuba City, he "did not even visit the Yuba to see the mine or the location of the mill. Indeed, the last half gallon of Rocky Bar whisky which he swallowed rendered such an excursion impossible. Stupefied and maddened by the fiery poison which he had imbibed in enormous quantities, he rushed off declaring the country worthless." An incensed Rocky Bar observer on November 1, 1867, after reporting this incident, continued with the admonition: "Gentlemen of the Alturas Mining Company, hurry and elect another president, and allow the present incumbent to subside into the position for which his tastes and talents best qualify him, viz: the presidency of a Pacific Street deadfall."

Disastrously for the mining operation, the inebriated Alturas Company president gave directions that the teams and wagons hauling the mill were to be sold on the spot. The mill itself was left to rust at Junction Bar, and the lumber for the buildings at Yuba City was abandoned to exposure and ruin during the hard winter which was at hand. The superintendent, J.W. O'Neal, an experienced California and Comstock miner, was to manage a useless enterprise. Great complaints against the evils of eastern companies issued from Rocky Bar, Atlanta, and Boise.

⚒ 🐎 ⚒

British investment, introduced to Atlanta by Matthew Graham, began to make an impact in 1868. Graham, who had started an arastra near Rocky Bar with a $300 grubstake during the South Boise gold rush, was a self-made man. Having advanced into the milling business at Rocky Bar, he next became interested in the Lucy Phillips at Atlanta. On the strength of his representation, the Lucy Phillips Gold and Silver Mining Company, Ltd., had organized in London on October 22, 1866, with £120,000 capital stock: 24,000 shares at £5 each. Within a year, three-fourths of the total capitalization had been subscribed. This subscription gave the company £90,000, a sum which was augmented by £72,000 on assessments collected that year (only £517 was listed as unpaid assessments, an indication that the stockholders were in earnest).

With something like $800,000 available by November 1, 1867, the Lucy Phillips was an enterprise of greater potential magnitude than the ordinary quartz mining venture in Idaho. Of the 301 stockholders, more than half (183) were British gentlemen, a class to which Matt Graham (the leading stockholder, with £4,000 to his credit for his interest in the mine) was graciously admitted. There were thirty-four clerks-in-order, and the others were from widely scattered occupations: a few surgeons, artists, merchants, attorneys, brokers, spinsters, and married women. In addition there were prominent figures in British society — two baronets, three rear admirals, and a number of military officers — and even some common folk, such as a leather cutter, a butcher, and a brewer. Development work on the Lucy Phillips had barely started before Matthew Graham left Yuba for London at the beginning of July 1867. Yet he had no trouble in arranging to get a large stamp mill authorized for the operation. The stockholders, in fact, were disappointed that the mill could not be brought in that winter.

As cheerful reports of development of the Lucy Phillips came to England during the winter, Graham was able to inform the stockholders in a London meeting on February 3, 1868, that a crosscut on their lode had been driven sixty feet without getting through the vein. The depth of their lode had not been determined, but for the moment, that was of no consequence. Enough ore

Atlanta (1877 or 1878)

already was in sight to supply their present machinery for a hundred years (this exaggeration suggests mostly the inadequacy of their machinery). Graham had not tried to get their new mill into the district that winter so as to save $5,000 in transportation and costs, but by the next summer the mill would be installed. By June, in fact, some twenty-five men were employed in erecting the mill, which, at ninety feet square, he could boast was the largest in Idaho.

Construction of the Lucy Phillips mill employed almost half of the fifty or sixty men at work in Atlanta by June 1868. But when the road could be reopened, so that the community was no longer isolated by winter snow, it was expected that activity would pick up. Charles Woodward had finished a new custom mill and was waiting only for some ore to work.

W. R. DeFrees had disposed of his interest in the Greenback mill, which had run a little the year before, but the new management planned to reopen it when the road became passable. As the situation worked out, the new owners seemed to have confined their work on the Greenback to developing the mine rather than running the mill. DeFrees now had acquired another property — 1,800 feet in the Leonora and 4,000 additional feet adjacent — along with the Northern Mining Company's ten-stamp mill, which he planned to put into use on his new mine. In the meantime, he set out for the East again to promote another big mill project to serve his new property. By now it was clear that although the Atlanta lode and many of the ledges

A cookhouse in Quartz Gulch served early day Monarch miners

Monarch mine in 1918

were predominately silver, they had enough gold to be mined and milled profitably without regard to the silver. Ore could be gotten out easily enough to process at $60 a ton, and with the mills then in use, much of it would return $100 a ton. The main drawback was that none of the mills was equipped to recover more than the gold. Tailings were saved so that they could be processed later for silver, but that kind of operation left much to be desired.

Expansion of the Lucy Phillips' holdings in Atlanta came on June 20, 1868, when Henry James, the company manager, purchased two hundred feet in another of Matthew Graham's properties. This deal apparently resulted from the prospect of serious trouble that ruptured Graham's previously excellent relations with the British company. Disquieting news came back to London that the title to the company's original mine was dubious. The Lucy Phillips' shareholders, in an extraordinary general meeting on August 13, 1868, voted to enlarge the capital stock of the company in an effort to get more funds by issuing new shares. Edward Bishop was appointed managing director for operations in Atlanta, and the company was authorized to purchase other mines or to operate a custom mill in the event no ore should be available from their own properties. In the meantime, the development of mining at Atlanta was set back seriously. An Alturas correspondent on July 13 indicated that

> the unfortunate trouble between Mat. Graham and the large English Company, has, for the time, thrown a wetblanket over operations in Yuba, but I am confident all will yet be made right, and our mines will prove that the curse of failure does not rest with them, but with the bad management or dishonesty of agents and others, who either know too little, or represent too much, and who are sent here without mining

experience, to do what only practical mining men ought to be trusted to execute.

Compensating, at least in part, for the Lucy Phillips setback was important capital investment from Indiana which boosted the development of Atlanta in the summer of 1868. The Monarch Company purchased 775 feet of the Atlanta lode for $45,000 in gold coin and $230,000 in stock. At least 175 feet of this purchase was regarded as singularly valuable; even by a crude milling and recovery process then available, $250 to $500 per ton could be produced. With better technology, this part of the lode would soon seem to offer prospects superior to the Poorman and Owyhee. The company pledged that a ten-stamp mill would be in operation by the fall of 1869, and that pledge was met.

At the same time, other activity kept up interest in Atlanta. Cyrus Jacobs continued to report good assays and new discoveries in his Minerva and prepared to get that property into production. Some milling was managed during the winter of 1868 and 1869 on DeFrees' Leonora. A ten-stamp mill processed 130 tons for a $9,000 recovery before the end of December, and because the mill was inferior (lacking pans, concentrators, and separators), the result was considered fine. Another $6,060 in gold came out of Leonora ore in January, with only half of the gold and none of the silver recovered. The Monarch started up that month too, handling a test run of six tons. Recovery of 1,200 pounds of dry amalgam — a rate of about $500 per ton — showed the possibilities of the Atlanta lode. Because the ore was almost entirely silver, efforts to recover silver as well as gold at last seemed to have started. Altogether Atlanta still had excellent prospects, and Matt Graham's fine discovery adjacent to the Monarch, in February 1869, was still more encouraging. The trouble was that no one had yet really solved the milling problem, although the Monarch seemed to be making headway. By the spring of 1869, the Monarch had begun to employ the Washoe process, and Atlanta was in a position to realize its early possibilities, providing that someone would get around to building a road.

During the spring and summer of 1869, efforts to expand production at Atlanta continued. The Monarch Company purchased the old Farnham mill, which had failed at Rocky Bar four years before, and later in the season acquired the Greenback mine in Atlanta for $300,000. Mining operations, however, continued to be limited. In the spring, only one mill was running, the Monarch on Atlanta ore. William R. DeFrees, after all his efforts to get the Leonora going, went broke early that spring. When he was unable to pay off a $3,000 note due April 9, his creditors attached the Leonora mine and mill.

Water was high that spring, roads were bad, and not much work was going on. Efforts on the part of Dr. Edward Bishop during the summer of 1869 to bring more English capital to Atlanta looked promising for a time, but did not lead to much in the end. Sent to straighten out the affairs of the Lucy Phillips, Bishop also represented the interests of the Gold Mining Company of Yuba, Ltd., that was incorporated July 30, 1869. An initial purchase arrangement in 1869 fell through, however, and little resulted from later investment efforts of this concern — other than to teach British capitalists to beware of mines in the United States.

A dismal conclusion likewise awaited the Lucy Phillips enterprise. Matthew Graham, who regarded himself as the agent to watch the British company's property, kept track of the elegant but not yet completed mill and other machinery from January 6 through November 30, 1869, at a rate of $10 per day. Realizing that he might have a hard time getting paid for his efforts, he filed suit on December 2 for $3,280. From this action he learned from the company that a £150 payment to him on August 30, 1869, was intended to represent a final settlement of all his claims against the Lucy Phillips concern. Graham managed to get a favorable

Large stacks of timbers were hauled into the Monarch mine

Monarch mill under construction

verdict in court. The Alturas County sheriff attached the Lucy Phillips mill and equipment for him on December 10, 1869, and the case, heard in the court of Chief Justice David Noggle, went in his favor. The luckless Lucy Phillips Company filed a report or two after that but showed little signs of activity otherwise. Eventually, everyone agreed that the company ought to be liquidated: a resolution of an extraordinary meeting of the stockholders "duly held at the London Tavern" resolved "that it has been proved to the satisfaction of the Company that it cannot by reason of its liabilities continue its business, and that it is advisable to wind up the same." An insurmountable obstacle then presented itself. Funds to hold a final meeting necessary to complete arrangements to dissolve the company could not be raised. Eventually, in 1880, the British Parliament came up with a simple procedure for getting rid of corporations as unsuccessful as this ill-fated enterprise. At last on August 31, 1883, the Lucy Phillips Gold and Silver Mining Company, Ltd., was formally dissolved. The Gold Mining Company of Yuba, Ltd., met the same fate on January 15, 1884.

Thus by the end of 1869, progress toward large-scale mining at Atlanta had made discouragingly little headway. Test milling of gold ore in 1867 and 1868, and of silver ore in 1869, had shown that the district ought to be developed. Some occasional arastra operations were still running at the end of 1869. Actual production, however, amounted to little more than some spasmodic gouging. Atlanta silver ore proved to be extremely refractory, and attempts to process it only to recover the gold did not turn out to be economical.

⚒   🐴   ⚒

Even though all major attempts to work the great Atlanta lode had failed by the end of 1869, promoters had one indisputable advantage: rich ore suitable for large-scale development was available in quantity sufficient to promise a fortune to an investor who might back a successful company. London interests wishing to recover British losses in the original Atlanta investments sent out William Nancarrow to examine the situation in 1870. Incredibly high-grade samples convinced him that the Monarch looked better than anything to be found on the Comstock. Yet terms of purchase were never worked out during several years of negotiation. Meanwhile, the Monarch continued to mill a small amount of high-grade rock for its Indiana owners, with its waterpower turbine running during the winter and natural hot water used to prevent freezing. Other properties were still less productive.

Really good properties were being held by their owners pending a solution to the recovery problem. Less promising prospects, however, found a ready sale to investors—some British and others from the East, who invariably met disappointment during those years. Such promotional tactics damaged the reputation of the Atlanta mines severely. A resident of Atlanta complained on March 22, 1871:

> This country has been infested with a set of 'bilks' as ever cursed any quartz country. They play themselves off on the hard working miners as capitalists who have, or control large sums of money, and bum their way into some wild cat, bogus quartz, then get a little provisions and tools, and leave the boys to work for them on jawbone, which is mighty current in this camp. They go on to New York and London, and try to foist their bogus stuff off on the market. These bilks bring our real resources into disrepute, and they have been a great drawback on the camp. Thank God, they have none of the good ledges in this camp, and they will not get them unless they can show something more substantial than jawbone.

Under these circumstances, only small operators gouging out a little of the highest grade ore were able to do anything at all. In summer of 1874, for example, a rich surface discovery revived the Monarch: 5,500 pounds of selected Monarch ore was packed out to Rocky Bar, freighted to Kelton, Utah, and finally shipped by rail to Salt Lake City. Purchased in Atlanta for $5,000, this small lot brought $11,000 when smelted. Warren Hussey, the buyer, thus made a comfortable profit. But that kind of operation could do little to develop the district.

The alternative—milling in Atlanta—was equally unsatisfactory. Recovery rates were so low that tailings processed a third time were yielding more than they had during the first two attempts. Even when Atlanta began to revive in 1874, Rossiter W. Raymond noted the extreme disadvantage under which local miners operated:

> The fact that nearly all the mine-owners are poor working men, unable to put their mines in condition for continuous working during the long and severe winters, the absence of facilities for transportation and of proper works for the reduction of silver-ores, together with the high prices for provisions and mining implements and materials, continues to be severely felt by the Atlanta people; but present indications prophesy that in

spite of the drawbacks better days are in store for the district.

Improved transportation and technology were the key to profitable, large-scale operations in Atlanta. Neither would be achieved without expanded capital investment in the district. In the summer of 1874, investors from Buffalo, New York, acquired a substantial extension of the discovery lode. (The original Atlanta discovery lode, aside from this extension hereafter known as the Buffalo, went entirely undeveloped until July 26, 1877, when Ralf Bledsoe finally started something more than surface assessment work on it.) With great effort, considering the lack of a decent road over the hill from Rocky Bar, the ten-stamp Buffalo mill was freighted and packed into Atlanta in 1876. At the same time, the Monarch (leased to some trustees for most of two years beginning January 1, 1876) began to make some steady high-grade production after six years of delay. Neither the Monarch nor the Buffalo mills, although equipped with the Washoe process for silver recovery, operated very efficiently at first. Their really high-grade ores (over $300 a ton) were still being packed out for shipment to Omaha or Newark. (In 1876, before the Buffalo mill was

Atlanta blacksmith shop

Buffalo mill (1877 or 1878)

installed, twenty-seven tons of that company's ore, taken to Newark at a cost of $900, yielded almost $19,000.) When the Buffalo mill started its trial run May 3, 1877, a 55 percent recovery was realized. Although this was far better than the twenty percent or so from the Atlanta mills in the past, the Buffalo management was still dissatisfied. A five-hearth furnace with a ten-ton daily capacity was added that summer. This improved recovery enough that high-grade Monarch, as well as Buffalo, ore was worked at least with partial success. At this point, Warren Hussey tried to sell the Monarch to San Francisco investors, but met with no success.

Atlanta had its early major building boom in 1876 when the Buffalo mill was being constructed and the Monarch leasors were beginning to operate. The two companies had sixty employees for construction and development primarily. These, together with quite a number of smaller gouging and prospecting enterprises, supported a community of about five hundred people. With Ralf Bledsoe's construction of a long-awaited Atlanta road from Rocky Bar in 1878, Atlanta was in a position to realize some of its early promise. Milling in Atlanta, while limited to high-grade rock, began to make a little headway after 1878. The extremely

high-grade Monarch and Buffalo ore that was being hauled out for smelting at least could go out over a wagon road. Moreover, by 1878 ore was being packed into Atlanta from Yankee Fork for processing in the Buffalo mill. Thus for several years, mainly from 1878 to 1884, the camp had considerable activity.

Yet the sad part of the story was that the costs of roasting and processing Atlanta ore were so great that only high-grade portions ($100-$300 a ton) of the lode proved to be worth milling locally. Renewed efforts by J.E. Clayton and other noted mining engineers did not begin to reach a solution for the large bodies of lower grade rock. (At this stage, Clayton and some of his associates reportedly arranged to sell the Monarch and Buffalo to August Belmont for $800,000; if they actually did so, nothing much seems to have come of the transaction.) During the early productive years before 1884, $500,000 was reported to have come from the Buffalo mill. Perhaps an additional $400,000 was milled in Atlanta by the Monarch. Yet fully $1 million of the $1.4 million in Monarch recovery for these years came from extremely high-grade ore (primarily 1,000 tons from the rich surface discovery of 1874) handled by the Omaha smelter.

Another fairly significant source of Atlanta production, the Tahoma, operated profitably for over two years after a $110,000 sale (following eight years of development) to investors in Meadville, Pennsylvania. But by the end of 1884 the Tahoma ceased to meet its payroll. Then, after going six months more in accumulating unpaid labor claims, the Tahoma default came out in the open when liens were filed for $18,000. Atlanta was plagued by a general economic collapse when one unpaid creditor sued another. Since the other mills also were grinding to a halt, the initial Atlanta boom came to an end. Even W.H. Pettit's new fifteen-stamp Monarch mill, which had commenced operation on December 15, 1884, ran only occasionally. By then most of the major Atlanta mines were bonded for sale again, because values less than $30 per ton could not be handled until improved recovery methods might reduce the cost of milling. By now, all available high-grade ore had been processed.

⚒   🐴   ⚒

Several years of effort to sell a combine of the major Atlanta properties to London interests came to a head when V.S. Anderson reported a $3.5 million deal with London capitalists early in 1891. This sale proved to be decidedly profitable to the original owners, who could not figure out a way to work their properties any further. But the Atlanta Gold and Silver Consolidated Mines, Ltd., found that even after considerable expenditure for development (presumably in excess of $150,000 beyond the $650,000 purchase), the recovery problem for low-grade ores could not be solved economically. After a final attempt with new funds raised in 1894, the British company had to search for another investor who might try to work the major Atlanta properties. To some it looked as if worked-out mines had been peddled to unsuspecting British capitalists. Yet by far the greatest production from the lodes the British purchased was still to come. While the British were searching for a suitable recovery process, the concentrating of Atlanta ores began on a limited scale in 1894, and the camp remained somewhat active until 1899. The Monarch fifteen-stamp mill resumed under new management for a time in 1902, and after another failure, the new owners undertook four years of Monarch development before installing a new electric-powered mill and tram in 1906. For the next several years the new company tried to figure out an economical recovery process which

would allow the new mill to commence work on Monarch and Buffalo ore. In the meantime, the Minerva, which had less refractory ore, resumed production with an electric-powered mill. The Tahoma and a property eventually known as the Boise-Rochester also produced for a time. By 1912 all were shut down again, awaiting an efficient recovery process.

Finally after the Boise-Rochester managed to run for two years after 1915, the Saint Joseph Lead Company purchased the property and eventually solved the recovery problem. Gaining access to the Monarch ores as well, the Saint Joe started up an amalgamation-flotation concentrator early in 1932. At this point, the problem of handling refractory Atlanta ore was solved after more than half a century of searching for an economical process. Modern production at Atlanta dates from that time. By far the greatest part of the $16 to $18 million Atlanta yield comes from the period of modern production after 1932. With construction of the Middle Fork road from Boise to Atlanta in 1938, another long awaited improvement in mining at Atlanta was realized, and the last of the old-time problems of that camp was solved.

Atlanta before 1900

BOISE RIVER

Compared with Boise Basin, placers along the Middle Fork and main course of the Boise River were harder to prospect, more difficult to process, and less productive after they were located and worked. Aside from their proximity to valuable basin properties, they might have gone unnoticed much longer than they did. Richer placers generally are deposited close to lodes from which they are eroded. Some finer particles of gold gradually float downstream where they eventually repose in sand and gravel bars up to hundreds of miles from their source. Some of the finest move on out into the Pacific Ocean where no practical recovery system has been devised.

Placers along the Boise River have been distributed in noncommercial quantities in valley bench lands as well as in more recent channel gravel. Occasional bars in bends along former channels, often high above more recent stream beds, contained enough fine gold to interest nineteenth century prospectors. They hardly could appreciate how much gold was involved, because a substantial part of these values came in particles too fine to be seen by the best microscopes available. Prospectors normally did not pack along high quality microscopes, and even if they had, they could not have processed such fine gold anyway.

An influx of hundreds of Middle Fork prospectors followed John Stanley's arrival in Idaho City after his party discovered gold at Atlanta early in August of 1863. Most returned disappointed, but a few were recovering from $8 to $12 a day in favored locations in September. Some spectacular placer discoveries on the Middle Fork of the Boise above Pfeiffer and Swanholm creeks came below Atlanta in the spring of 1864. Although that area was eventually dredged, it showed little early promise.

Not later than the summer of 1864, river placers above Boise were gaining recognition in the area where Lucky Peak Dam was completed in 1952. Additional discoveries at Boise King (near Pfeiffer and Swanholm creeks) followed in June 1867, and an early hydraulic ditch from Black Warrior Creek served the area. Other placers above Mores Creek and above the South Fork at Cottonwood Creek were found in the summer of 1869. Mining was under way near Twin Springs eight years later; after two more decades, a large siphon project supplied water for a massive hydraulic operation good for $150,000 on some high bars there. Large-scale Chinese operations high on the ridges near Black Warrior also contributed to Middle Fork production, and a Black Warrior lode location in August 1903 led to considerable additional activity

there from 1904 to 1906. After Black Warrior declined, Arthur W. Stevens reopened about 480 acres of Boise King placers, which he estimated as carrying placer values at forty feet of depth. New cabins were built, and ditches were reopened and enlarged in 1909. By 1912 he had a large flume to operate a hydraulic giant as well. But his placers lacked sufficient value to support that kind of operation. Dredging of $200,000 during 1940 to 1942 and 1946 above Boise King accounted for most of the remaining Boise River production. Activity resumed in 1981, with a bulldozer and backhoe to supply placer gravel for a 1982 recovery program.

VOLCANO

Widespread prospecting ranged over much of southern Idaho in 1864. New finds turned up in several scattered places, although the earliest, while promising in the beginning, did not really amount to anything in the long run.

Volcano was the name bestowed upon a new lode district found March 12, 1864, on the eastern extension of Bennett Mountain between the Camas Prairie and the Snake River Plain. At the close of the 1864 season, twenty-five to thirty veins were known there. Some assayed at $180 a ton, and one reported by T.A. Patterson ran as high as $2,100. Much effort went into prospecting, but practically nothing into development. The miners preferred at that time to wait for a stamp mill to test their properties. The 1865 season, therefore, was a very quiet one. At its end, W.R. DeFrees' ten-stamp mill finally arrived. Arrangements were made to run the mill all winter, even to the extent of bringing in enough hay to supply horses required to haul ore to the mill. A tunnel was started to strike the five-foot vein of ore for the operation. With surface assays of $600, DeFrees expected a handsome profit. Not until September 1866 did DeFrees' company finally manage to assemble 100 tons of rock for a test run. DeFrees was greatly disappointed to find his recovery process returned only about $10 to the ton. Operations naturally had to shut down altogether when the rich outcrops turned out not to be ore at all, at least for that mill.

Copper mining at Volcano began to attract interest when William Clemens went to work in 1900. After 1924, a 1,453-foot tunnel was driven before work shut down in 1931. Now, after 118 years of sporadic attempts to develop major mineral production at Volcano, an open-pit silver-copper mine began operating in 1982. With eight miners at Volcano and thirty more employees at a $500,000 360-ton concentrator in Mountain Home, the mining venture began processing a 3,000-ton ore reserve at an initial rate of 80 tons a day. Over $1 million has been invested in this latest development at Volcano. At $10 an ounce for silver and 70 cents a pound for copper, ore averaging $260 a ton is expected to return a $3 million annual profit. So Volcano, like some other early discoveries in Idaho, has eventually begun to justify the long-held optimism for its potential—but only after transportation and technology far different from anything envisioned in 1864 finally became available.

LITTLE SMOKY

After John Stanley's prospectors found gold around Atlanta, another expedition set out early in April 1864 to check out his modest gold discoveries in Stanley Basin. Of all the mineral country this group examined, Little Smoky eventually proved most important. Heading from Boise up past the South Boise mines to the upper South Fork, these prospectors met with at least limited success on Little Smoky before crossing north into Stanley Basin in pursuit of other excitements. Eventually the expedition's leader returned to Little Smoky at the time of the rush to Wood River. In the meantime, extremely small-scale yet successful placering had gone on there since 1873. Finally, quartz discoveries in 1879 led to the development of a district that produced enough in 1886 to cover $15,000 to $25,000 in operating costs, pay for a $15,000 concentrator, and provide $30,000 in dividends. Late in 1886, Salt Lake City investors purchased a major group of Little Smoky lode properties for $105,000 in cash. Eventually these lodes were credited with a $1.2 million production. Gold accounted for about $200,000 of that amount.

Little Smoky placers continued on a modest scale as well. Newton Rives, who came to Idaho in 1861, worked his claims all by himself for more than two decades after 1873. Chinese miners purchased some Little Smoky claims in 1895 and expanded the placer operations. But none of these compared with the lode mining. Production was increased in 1980 with the reprocessing of old dumps at Carrietown.

BANNER

Silver lodes in Banner were traced out of already active Crooked River placers, which in turn had been found by prospectors radiating from Boise Basin in 1863. The placers, known as the Rocker

Diggings, had enough silver with the gold that two of the operators there concluded in June, 1864, that a silver lode ought to be sought in that locality. Inexperienced in such prospecting, they came over to Placerville and induced Jess Bradford, an old hand in dealing with silver ledges, to come to examine their part of the country. Bradford persuaded James Carr to join him, and after a few days prospecting northward from the Crooked River placers, Bradford and Carr found the Banner ledge on July 6, 1864. Taking some samples, they returned to Placerville only to find that the current craze was north on the Payette. The day after their return, Bradford joined James H. Hawley in the Payette rush; Carr, who was a partner with Bradford and some others, remained behind, however. After some time, he got the samples from Banner assayed. They proved to be rich enough in silver to stir up considerable excitement, and after waiting about as long as he dared for Bradford and Hawley to return, Carr located the Banner lode on August 8, 1864. In doing so, not only did he list the names of the actual discoverers, but he also added to them the names of Hawley, the rest of their partners, the assayer, and one or two others who promised to be useful in the venture, including one who offered to bring in a stamp mill if he could have an interest in the claim. As news of the high assays began to get out, interest picked up and on August 23-24, 1864, about two hundred prospectors rushed to Banner.

Within a few weeks, the two new towns of Banner and Eureka were rising, and about fifty new ledges were discovered. Aside from the original Banner, no one knew whether any of the rest of the fifty had any value or not. The ledges ranged in width from two to eight feet. Excitement among the 120 miners who had remained there reminded Morton M. McCarver, a prominent resident of Oregon who had seen a lot of mining rushes, of the early days on the Comstock at Washoe. In the commotion, the story of Banner was already a little confused. According to the way McCarver had heard it in September, the ledge had been found by parties searching in the wrong place for an improved route from Placerville to Rocky Bar. Other than that, McCarver's story matched that of James Carr, one of the discoverers. Before winter closed in, the pioneers of Banner were ready to go to work. In spite of severe weather, much development had started on the new lodes.

By the end of February 1865, a tunnel toward the Banner vein progressed 270 of the 300 feet that were expected to be necessary, and eight other tunnels, ranging mostly from 100 to 160 feet (one was 240), were being driven on other properties.

Banner

The community was cut off from all supplies by snow, which was eight feet deep in town then. But enthusiasm for the mines was increasing rapidly. E.J. Neale reported that prices for the properties were doubling—presumably only in the estimation of the owners, for no one could get in to buy. Two of the original locators, James Carr and I.L. Tiner of Placerville, were putting up a store, but no supplies or merchandise would be for sale until spring. A long hard winter had delayed further development considerably, but by mid-August, the Banner tunnel penetrated the required 300 feet, where the vein was struck 150 feet below the surface. Then on October 30, at the end of the season, a ton of Banner samples, collected for eastern promotion, left Boise for New York.

Arastra production came to Banner in the early days, but stamp milling took longer to get under way. The lack of a road held the camp back until 1868. By mid-summer that year, a shaft had been sunk on the Banner vein for 80 feet below the end of the 300-foot tunnel, and an arastra was kept busy crushing the ore that was being recovered. After a decade of arastra production, G.W. Craft finally got a mill into Banner in 1874, and capital from Elmira, New York, helped to develop the district still more in 1878. Stamp-mill production continued at Banner for more than ten years, with a considerable spurt of activity there from 1882 to 1884. Then a $400,000 sale came in 1884. Before the district shut down in 1921, production of silver totaled nearly $3 million.

BOISE RIDGE

Prospecting on Boise Ridge not far from the Idaho City road disclosed some promising lodes in the winter of 1864-1865. Additional discoveries followed periodically, leading to K.P. Plowman's location of a productive Shaw Mountain mine on Deer Creek in May 1877. After utilizing an arastra there, he brought in a mill during 1881. Another substantial 1877 discovery in McRay's gap adjacent to the summit of the Idaho City road accounted for the installation of a ten-stamp mill there in 1879. In 1894, an additional stamp mill still closer to Boise began to crush ten to twelve tons of ore a day. James Flanagan employed fifteen miners in his mine and mill, located only five miles from Boise.

After an interesting lode discovery in 1891, the Black Hornet on the east side of Lucky Peak gained considerable publicity. Denver investors showed interest in the property in the summer of 1895, and finally a $100,000 purchase was completed in the fall. Daily carloads of $40 ore were shipped to Salt Lake City that fall, and a $69.65 nugget found two years later helped promote the district. By 1896 a group of Black Hornet mines had produced about $30,000 with $24,000 of that amount in 1895-1896. Another $400,000 followed later. Farther west, small lode developments on Cottonwood Gulch above Rocky Canyon added slightly to Boise Ridge production. Even though Boise Ridge did not begin to compare with Idaho's major gold producers, mining on Shaw Mountain contributed in a modest way to the state's mineral wealth.

HAILEY GOLD BELT

Although its location only ten miles from Hailey places it close to a large district of Wood River lead-silver mines, the Hailey Gold Belt had quite a different development as an independent district. Eight prospectors from Rocky Bar located two lodes, Big Camas and Black Cinder, on September 11, 1865. Two of the miners eventually settled in Broadford and Bellevue, but the development of these properties was delayed until after the 1880 rush to Wood River made work more feasible by bringing miners, supply facilities, and more convenient transportation to nearby camps.

By the summer of 1883, a 450-foot tunnel had been driven to crosscut the Camas lode without actually reaching it. At that point, mining of the outcrop was chosen as a way to get started. This proved to be relatively simple at first: an 800-foot vein of $15 ore projected several feet above the ground over a width up to fifty feet. So surface operations began more as a quarry than a mine. Then an investor from Saint Joseph, Missouri, bought a major property for $10,000 in the summer of 1884. Milling began June 1, 1885, with a ten-stamp outfit that soon was doubled in capacity. An

initial cleanup yielded only $605 on June 17, but
within two years, a twenty-stamp mill was grinding
out concentrates of $40 to $350 a ton for shipment,
in addition to free gold recovered at the site.

To improve shipment facilities for concentrates,
a Hailey, Gold Belt, and Western Railway company
was organized June 16, 1887, and a line was sur-
veyed immediately. More than surveys were
needed, though, to offer much service to the
district. A new town of Gold Belt was started
anyway, complete with a saloon and restaurant,
blacksmith shop, carpenter shop, livery stable, and
other conveniences. A decade later, recovery from
the Hailey Gold Belt was increased by the introduction of a cyanide process on October 1, 1897.
Jerome B. Frank of Denver succeeded eventually in
reworking all the tailings from the Camas
concentrator, raising production to 102,000 ounces
of gold.

LEESBURG

Mining at Leesburg grew out of mining in Montana, which had formed an 1862 extension of the
rush to Florence. Indeed, until the summer of 1867,
the Leesburg placers might as well have been a part
of Montana for practical purposes. Then direct
routes from Boise and Idaho City to Leesburg
changed the pattern of traffic and supply.
Prospectors fanned out from Leesburg in all
directions. The wild country of central Idaho was
examined more thoroughly than had been possible
before. By 1867, a limited amount of mining was
going on around Stanley, which soon had a store,
and in 1868 a hydraulic giant was in production at
Robinson Bar. In the summer of 1869, a rush to
Loon Creek, a direct outgrowth of the Leesburg
expansion, brought an additional wave of
excitement into that part of the Salmon River
Mountains. The future major discoveries
throughout the rest of the neighboring mining
country—Yankee Fork, Bay Horse, and Clayton in
particular—were a matter of only a short time.
Yankee Fork, in fact, had its beginning in 1870, and
the others only a few years later. Aside from leading
directly to the founding of an important service
community in Salmon, the rush to Leesburg was a
major factor in bringing mining to a large,
undeveloped area of central Idaho.

Led by F.B. Sharkey, a five-man discovery
party—one of whom had mined at Pierce and
Florence—left Deer Lodge, Montana, on June 10,

Leesburg placers

1866, to examine the country to the south. Reaching
Napias Creek on July 16, the explorers sank a prospect hole to bedrock and found high-paying gravel.
Organizing themselves as the discovery company,
they mined profitably in Leesburg Basin for a few
years. After that, at least three of the five became
prominent cattlemen, two in the Lemhi country
and the other at Grangeville. These men had come
to build the country in which they found gold.
Sharkey, their leader, along with the others who
went into the livestock business, spent the rest of
their lives in the region.

News of the Lemhi placers reached Bannock,
then the capital of Montana, in time to start a rush
to Leesburg on August 19. S.F. Dunlap, the regular
Bannock correspondent to the *Montana Post*, a
Virginia City publication, gave a brief account of
the excitement:

> . . .our people have been pursuing the even tenor
> of their way, and we were quite unprepared for
> the stampede that set in yesterday for the Lemhi
> Valley—the tide still rushes on, and unless there
> is a counter current in a few days our streets will
> be almost deserted. The valley in which the new
> placer diggings have been discovered is very
> extensive, and it is said, that it resembles the
> Boise Basin. The diggings are shallow and the
> dirt prospects evenly, from seven to thirty cents
> to the pan on the bedrock. What the results will
> be, we cannot tell, but it is certain that the
> excitement is great, and the faith of the stam-
> peders strong.

But ten days later, he reported: "The Lemhi stampede is at an end. It turns out that the new diggings
will not pay much over wages, yet some are preparing to mine and winter there, but most have returned to Bannack to pursue their legitimate business interests." Then on September 6, he modified
his pessimistic account of the prospects of Leesburg:
"It appears that there will yet be quite a camp at the
new diggings on Lemhi. Quite a number are mining
there and the dust that is brought into Bannack for

goods is as fine as our own." Again, on September 17, he wrote more hopefully: "Trade is brisk, and scarcely a day passes but a pack train comes in for goods for the Lemhi mines. We believe there will be a good camp there next season. There can be no reasonable doubt about it." Finally, on November 28, Dunlap noted the beginnings of Salmon City and the bright prospects for Leesburg:

> Bannack still progresses. Pack train after pack train from Salmon River mines arrive in town for goods, and give the best proof in the world that the above mines are good and extensive. It is said that the 'dust' assays as well as the Bannack; and five or six hundred miners are preparing to winter at the mines, while others are building their cabins at the crossing of Lemhi Valley, fifteen miles from the mines. The valley is warm, and good for winter grazing. We believe Salmon river will be the best mining camp in the Rocky Mountains next season.

By that time, Leesburg had gone through the rather ordinary pattern of a gold rush: the stampede to a new district, followed by the initial disappointment of a substantial number of fortune hunters, followed in turn by the establishment of a permanent camp of prospectors and miners who begin to find evidence that their mines have a great future. As was usual with a discovery in midsummer or later, prospecting did not get far enough before the mining season closed for anyone to have much of an idea how extensive the new district would turn out to be.

Construction at Leesburg made considerable headway during the winter. Named for the Confederate General Robert E. Lee, the new community, like a great many Idaho and Montana mines at the end of the Civil War, was composed of strong Confederate sympathy. Some northerners also inhabited the Lemhi mines, and in the middle of November, one of their number started Grantsville right next to Leesburg. The two towns, in fact, were really one. The pretense of Grantsville was kept up for a while, but Leesburg was the name

Leesburg in 1870

that survived. A Grantsville correspondent on December 18, 1866, in writing to the *Montana Post* noted the progress of the community in that forty buildings had been completed, among which five stores, three butcher shops, a blacksmith shop, and a feed stable already were in operation. Five hundred miners already had arrived in the winter, and the number was rising. Not too many claims had opened the preceding fall, but the ones that had got started paid some $12 to $16 a day. Some men were still prospecting in December, "but the most are engaged in building on speculation, on their town lots." As the winter wore on, prospecting became impossible, so cabin building was about the only thing left for anyone to do.

For the extent of mining operations and discoveries that first fall, James Henity, writing from Leesburg on January 2, 1867, gave a clear picture:

> I will give you my opinion of this camp. I think it will be the best and largest camp ever struck in the country outside of Boise Basin. There have been no very large strikes made yet, but the country as far as prospected thoroughly will pay well. We have eight large districts, two of which have been struck since Sam left. These diggings were struck late in the fall and there have been no merchants or grub here, so men have had no time to prospect until after snow fell. The gold is of a course order and runs nineteen dollars and seventy cents per ounce. The largest piece was picked up in Bear Track District weighing $15.75. These diggings are seven to ten feet deep, with three feet of gravel. Prospects are from five cents to three bits to the pan. Raft Creek has been located from its head to its mouth. I saw thirty-six dollars that was panned out of one prospect hole, nice coarse gold, at an average of four bits to the pan. Water is plenty. In Jump's district, the creeks and bars prospect well for five or six miles. George Byron is rocking there and making seven dollars a day. Petot's district prospects for four miles. Idaho district, a continuation of this creek, will pay as far down as has been prospected. This district is called Napias, and is the first discovery, and here most work has been done. From five to sixty dollars to the hand has been made. There are three districts on the other side of the divide that prospect well. Everything here is on the move. Twenty-six houses have been built in town. There are six stores, two butcher shops, and about a hundred more houses going up. There is a large store and saloon at the river. Weather pleasant; snow two feet deep. Flour $22 per hundred; beef 17 cents per pound; bacon 60; coffee 90; sugar 60; beans 50; potatoes 20. Mining tools and clothing high. Some very rich quartz has been struck, but nothing much will be said or done about it until next spring.

Through the early part of January 1867, pack trains continued to go back and forth to Leesburg.

But as winter snow piled up, communication was limited strictly to men on snowshoes, and it was not too often that anyone was bold enough to make the trip in or out. Logs for cabins could scarcely be procured as winter advanced, although construction on J. Marion More's 3½-mile ditch continued to progress. Then on January 26, when snow depths had reached four feet in town and ten to twenty feet on the trail, the restless miners of Leesburg managed to go on an eight-mile stampede to Mackinaw Gulch, the big discovery of the winter. There, under eighteen inches to two feet of black surface muck, prospectors had found gravel to the depth of six feet which ran rather evenly at one to three cents a pan. Nothing much could be done to develop these new placers during the winter, but they did bring new hope to a camp at a time when hope was needed.

By the beginning of February, Leesburg was running short of supplies, and the stock of fresh meat was exhausted. A correspondent to the *Montana Post* suggested some other needs as well, particularly medical help: "A doctor is sadly needed here. We have two men with ax cuts in their feet; two frozen men; one gunshot wound, and quantity of coughs and colds, together with some rheumatism, and there is no physician nor a particle of medicine in the camp." With prices for bacon and other essential commodities rising, a twelve-man shoveling company was organized at the beginning of February to clear the route out to Salmon.

The plan was to operate the trail as a toll enterprise from the time the trail could be opened until the snow melted. The project was expected originally to take about two weeks. But with new snow adding to the problem the shovelers were at it the entire month of February and a week into March before they finally succeeded in clearing the way for a packtrain on March 8, 1867. Two weeks before the route was cleared, Leesburg had run out of bacon. When eight head of cattle at last were driven in through a narrow channel cut through the twenty-foot drifts and generally deep snow on the trail, the procession was certainly "a welcome sight to the denizens of this. . .isolated but rising town." Toll rates were seventy-five cents an animal each way, and for once no one complained about paying the toll. Considering that the minimum depth of snow on most of the trail was reported to be five feet, shoveling out an eighteen-mile slot wide enough for pack animals to get through had involved hard work. Perhaps the length of the deep snow stretch was exaggerated in the accounts; there is good reason to believe that the entire population of Leesburg could hardly have shoveled that much

snow before midsummer. Nevertheless, the Leesburg toll pack-trail was certainly one of the exceptional ventures in the history of mining transportation.

⚒   🐴   ⚒

The excitement over the Lemhi mines continued to build during the winter, and by the end of April, the population had risen to about two thousand. More miners were on the way, from places as distant as California, although continuing snowfall prevented further prospecting or mining. Yet once the mining season could open, plenty of country would be available to prospect. William Clemens, a prominent Idaho miner for a great many years, reported from Leesburg on April 28 that the mining

Leesburg ditches, flumes, and placers

Leesburg

country there looked even more extensive than Boise Basin. Some rich prospects had been found and much of the country was known to contain gold. The problem was to figure out just how much of the gold region would prove rich enough to work.

Stage service from Montana over Lemhi Summit to Salmon and Leesburg was available for those who joined the Lemhi rush, but the service and accommodations along the road were certainly primitive. Ben R. Ditles of the *Montana Post*, who arrived in Leesburg on April 29, described his overnight stay in Salmon, amid a great stir and bustle of construction, as a memorable one:

> The small hotel, now the only place of resort for travelers, is crowded to overflowing, and a look into the large front room, on the night of my arrival, would have disclosed to some readers a strange sight. Some twenty men — some bound for the mines, others bound for Montana — all stretched out in their blankets with a goodly admixture of dogs who, every now and then, would become quarrelsome, and thereby keep the sleepy travelers awake. But all this was soon changed, a fine, commodious hotel will soon supersede the small affair of today, and Salmon River [he means City] will outrank any of her sister mining towns in Montana, the central point of supply for the mines.

From Salmon, which had definitely surpassed Lemhi City as the community that served those on the way to the new mines, the trip to Leesburg shaped up as a long hike over the snow-cleared trail. By the beginning of May, those coming in about balanced those leaving, and the population of the Lemhi mines stabilized at about two thousand.

Held back by a decidedly late season, a limited amount of mining around Leesburg got under way by the middle of May 1867. By then, it had stopped snowing for the most part, and the great shortage of provisions had been supplemented by an excess which caused a market collapse. Merchants from Helena, Boise, Idaho City, and Walla Walla who had rushed in with entirely too many supplies now faced a price failure that benefited those who had lost everything trying to survive in Leesburg that winter. Moreover, three to four new pack trains were arriving daily. The Leesburg building boom still was on, with 130 houses in various stages of construction, all for sale at prices ranging from $3,000 to $5,000. One or two mines were for sale also. Mose Milner of Walla Walla sold a claim for $5,000, but that was the only big transaction. All that was needed to make Leesburg into a fine successful camp was mining activity. Only a few properties, which had been in production the previous fall, could be worked before June. In May the editor of the *Montana Post*, who was then in Leesburg, reported that "hundreds of broken men now lie here awaiting the disappearance of the snow."

There was abundant country to prospect around Leesburg. As the snow receded, hundreds of men went out, and "excitements" were the order of the day. A particularly alluring one took about half the population on a wild chase eighty miles toward Warren on May 22. Then another rush to the head of Little Lost River on June 15-16 attracted a multitude of Leesburg miners. From the gulches actually open, production figures were promising, but not spectacular. When Anderson Cox of Walla Walla departed from Leesburg on May 22 after several months stay, he reported that "no big strikes or very rich claims had been heard of." The best of them were good for $12 a day. J. Marion More, as usual, had gone in and dug a big ditch. For that particular season, however, the problem seemed to be too much water, and it was hard to say whether his ditch was going to be necessary for awhile.

Leesburg remained a relatively placid camp as the 1867 season got under way. Mining had been "held but in moderate estimation by the mass of those engaged" in opening the district in the fall of 1867. Then Leesburg had "flashed into a meteoric brilliancy during the winter months when mining and prospecting were impractical, and had lured thousands from good wages and paying diggings." When prospecting resumed in the spring, a great disillusionment set in that was characteristic of many such mining rushes. A sober industrious camp grew up there, however. J.C. Bryant described Leesburg as anything but the wild community that was thought to result from a gold rush: The town with 150 to 200 houses "of the style and kind usually built in new mining camps. . .is unusually quiet for a mining camp and free from fights and disturbances. Everyone who works appears to be there for business and not pleasure. There is very little gambling and comparatively little drinking and spreeing such as generally characterizes a new camp." This sobriety he regarded as "contrary to all precedent."

An unfavorable season in 1867 set Leesburg back more than those who had worked so hard to build the town anticipated. Part of the trouble arose from the need in the northern mines for a big new discovery to attract the surplus prospectors left without much to do at that time — and Leesburg just happened to spring up when such excitement was needed. The placers did not justify the magnitude of the winter rush; in a situation, however, where great pressure existed for a big new discovery, Leesburg served the purpose. By the summer of 1867, it seemed to the compilers of the Idaho section of the annual government report on the mineral resources of the United States west of the Rockies that Leesburg had been considerably overestimated. In a brief statement, the report noted:

Last fall some mines were found on branches of

Salmon River, not far from Fort Lemhi. Exaggerated reports of their richness caused quite an excitement. The probability is the reports were circulated for the purpose of selling claims. It is said that one claim offered for sale prospected well in the snow above the earth. Accounts are conflicting as to the value of the discoveries, but all agree that there are some half-dozen claims on each of four or five gulches that will pay well. Some assert that these are all; others maintain that Lemhi abounds in extensive placers which will yield $5 per day to the hand, though it is generally conceded that they will not justify working at present, except in a few of the gulches.

Perhaps the most careful evaluation of the Lemhi mines was that of Cyrus Jacobs, who had observed the situation for a month when mining was getting under way in June. He concluded that an unfortunate season in 1867, retarded by late snow and thereafter depressed by excessive high water which made mining difficult, would be followed by an improved one in 1868. An army of prospectors was available in 1867 to examine that whole part of the country. Jacobs felt that they ought to be able to open enough placers to bring out a better return for the next year. Production figures for Leesburg — $250,000 for 1867 and $750,000 for 1868 — confirmed his analysis.

Miners leaving Leesburg in July 1867 (and by then more were leaving than arriving) came away with the impression that while the area certainly

Cy Jacobs

had some good productive claims, all too many of the placers were better designated for Chinese than for American miners. They also complained that too many of the placers were ruined by large rocks and boulders which impeded possible recovery. This kind of reaction was typical of the discouragement that followed practically any gold rush. After the surplus population cleared out, Leesburg settled down to the reasonable production rate that the original prospectors had expected.

During the 1868 season, Leesburg had no idle men in camp, and for those who stayed, the mines paid pretty well. A company on Sierra Gulch was "making money fast" with nine strings of sluices; a bedrock flume on Napias Creek promised good results. Unfortunately, a catastrophe to the second Napias Creek bedrock flume set the district back severely. After a year's work and $2,000 had been invested in installing it, the flume was destroyed when it filled with large boulders washed in from an unexpected collapse of a feeder tunnel. The year's work on the flume was regarded as just about a complete loss. Nevertheless, production for 1868 was triple that of 1867, and Leesburg at last showed signs of amounting to something.

Even at the end of 1868 a reliable observer, E. T. Beatty of Salmon, felt that the Lemhi mines around Leesburg were still barely opened. Much ground worth $4 to $6 a day still was untouched. Having no way of anticipating the future distraction in the rush to Loon Creek, Beatty felt that 1869 should suffice to bring the area into full production.

Although the suggestion that Leesburg had more gold than Boise Basin proved to be entirely unfounded, mining on a more modest scale continued around Leesburg for years. Yet, the 1870 population for Leesburg of 180 suggests that during the first winter, when most of the inhabitants could find little to do except construct houses and stores, they had overbuilt the town substantially.

Later lode discoveries helped maintain interest in Leesburg, but most gold values came from early placers. Before too long, Leesburg's placers interested Chinese miners mostly, and whites generally headed to other camps after their early work had recovered most of Leesburg's wealth. Prospectors soon noticed a large number of other mining possibilities in Lemhi County. For that reason, mining at Leesburg accounted for considerably more than $6.25 million which came from local placers. By 1895, Leesburg had only four or five whites and about fifty Chinese. But several neighboring camps

which grew out of Leesburg's activity continued to thrive.

Lode developers at Leesburg had identified some massive low-grade veins to exploit, but values were limited enough that they did not attempt to install large mills. O. E. Kirkpatrick, who walked 92 miles from Red Rock, Montana, to Leesburg in 1898, spent $300,000 in forty years developing a lode there. He had a ten-stamp mill in operation within five years, and kept up production when economic conditions were favorable. After a $20,000 purchase of a promising lode by Nebraska investors, another ten-stamp mill was set up at Leesburg for a time. Neither of these mills began to rival the output from Leesburg's early placers, but they occasionally helped sustain a declining camp.

Most of Leesburg's early placers had been shallow and were easily worked. But some large placer deposits could not be handled at all without extensive ditch and drainage systems. By 1908 a seventeen-mile ditch and flume system was available to facilitate production of Leesburg's major untouched placers. More than fifteen years of inactivity followed. Finally in 1926, hydraulic giants went into production. These shut down after a couple of seasons, and Leesburg placers remained dormant until 1934, when increased gold prices and low mining costs spurred a brief revival. A dragline operation in 1939 brought more life to Leesburg, but gold mining everywhere was shut down in 1942, and Leesburg faced more inactivity.

In 1982, after a year's experimentation with a less effective gold recovery system, Leesburg gravels (which, with increased gold prices, ran $15 per cubic yard) came back into production. A mining outfit, using earth-moving equipment capable of handling 2,000 yards a day, began processing a large volume of placer deposits that had been uneconomical to work in previous years. Aside from this renewed mining activity, only a few old cabins and traces of early work remain as a reminder of Leesburg's nineteenth century past.

LEMHI

Aside from Leesburg, more than a half dozen substantial mining districts in the Salmon area gained attention within a decade of intensive prospecting after 1867. Along with later discoveries at Gibbonsville, Shoup, Agency Creek, and Ulysses, these constitute one of Idaho's two largest blocks of mining districts. (Wood River-Sawtooth-Copper Basin-Mackay forms the other: both are similar in area.) A promising lode discovery on Comet Creek north of Salmon accounted for much of the initial

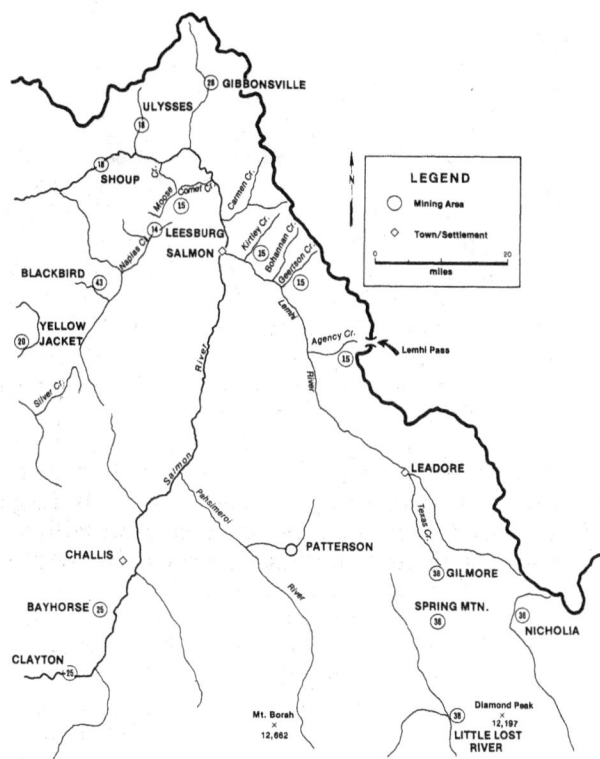

Lemhi mines. Numbers in circles refer to index map of mining areas, page 2.

excitement in June 1867. An interesting stampede to an unknown destination, which was finally revealed as Comet Creek a month or so later, created quite a stir in Idaho mining circles.

Placer prospects in a number of streams along the Salmon also commanded attention, although an excessive number of rocks and boulders reduced their value to $3 a day. These placers soon were disparaged as China diggings, and many of those who had rushed several hundred miles to examine them returned home "weary and disgusted, if not disappointed." On July 7, 1867, Salmon's *Semi-Weekly Mining News* "still speaks encouragingly of the prospects along the Salmon river [according to the *Idaho Statesman*, July 23] but we should judge from the tone that its faith was rather weak." Such press reaction from outside the area was typical: most newspapers discounted gold discoveries reported for other regions. In this case, although extensive prospecting had been going on there for several years, nothing too consequential had been found aside from John Adams and E.P. McCurdy's Comet Creek lode. The reaction of J. Marion More, who went out to examine the situation, was reported in Boise on August 7:

One very rich quartz ledge has been found. The placer mines he [More] does not think very valuable, judging from the reports and from the fact that prospecting has been done there for four or five years past without flattering results.

More's suspicion had a good foundation, although a number of placer districts did materialize along the river.

With an 800-foot-long surface exposure, the new Comet lode had enough good small assays to gain the slightly inflated reputation of having an outcrop richer than any others yet discovered in Idaho or Montana. Ben Anderson was willing to pay $300 for two hundred feet of that lode without bothering with further prospecting. Then, the richest outcrop known in Idaho or Montana drifted into obscurity. Such transactions were not noted for their rarity during Idaho's gold rush years. Yet prospecting continued. A slow season at Leesburg encouraged miners to examine a wider area, and other less important camps gradually emerged.

⚒   🐴   ⚒

In less than a decade, E.T. Beatty (an old Rocky Bar miner who had settled in Salmon) reported on January 26, 1876, that two arastras were producing gold on Kirtley Creek, that another arastra near Leesburg was paying well on Arnott Creek, that a six-stamp mill was active on the west side of the Salmon at Carmen Creek, and that New York investors had a ten-stamp mill headed for Geertson Creek. Placers also gradually came into production on Bohannon Bar about eight miles above Salmon. Because water was available for a short season each year, this six-mile-long bar (and a half mile wide) could only be worked slowly. In about two decades of seasonal operation, more than a $500,000 worth of $18 an ounce gold was recovered by the time Idaho became a state in 1890. By then, only about a tenth of Bohannon Bar had been processed. Surveys for a forty-mile Lemhi ditch (complete with two or three flumes) provided a reasonable, yet expensive, solution to this water problem. After dredging became practical, Kirtley Creek and Bohannon Creek placers became more productive.

Not all of these mines needed, or could get, supplementary water from the Lemhi or Salmon rivers. Moose Creek basin, north of Leesburg basin, provided $500,000 yield to David McNutt, who had enough water at high elevation to operate a six-inch hydraulic giant and a bedrock flume over a more extended annual season. On the Moose Creek-Salmon River divide, an extensive field of rich float boulders (successor to the Comet discovery of 1867) furnished ore for a five-stamp mill on Moose Creek.

E. T. Beatty

Dredge on Kirtley Creek (1912)

Bohannon dredge (1912)

Work went on for many years, with a revival during the Depression when gold mining paid better. Around $400,000 came from that unusual assembly of lode material.

An interesting new direction in the Lemhi mines came with F.B. Sharkey's location of the Copper Queen on Agency Creek in 1883. Although less important than a number of other copper mines in Idaho, it produced about $100,000 by 1910. While this was under way, Bohannon Bar production increased to around $200,000 in spite of an early dredge failure. A dredge on Kirtley Creek did better after 1908; production there may have totaled $1.2 million. These operations by no means have exhausted the area's mining possibilities: in Lemhi Pass a large thorium deposit—much of which occurs in Montana—has been explored but has remained unproductive until 1980.

PEARL

Although development of lode mining around Pearl did not come until much later, a considerable stir was created in Boise on December 7, 1867, when the proprietor of the Dry Creek station and ranch showed up with some good looking quartz specimens from two substantial veins on Willow Creek. The ledges, two hundred feet apart, were parallel and well defined. The gulch between them

prospected well, and miners hoped to be able to recover $8 a day by putting in a reservoir to provide water for their operations. At the very beginning of the excitement, the sale of a hundred feet on one of the veins for $400 showed that there was interest in Willow Creek properties. But after a limited amount of work was done at the Red Warrior at Pearl in 1870, the district remained dormant until gold mining staged a comeback during the Panic of 1893. While Pearl was inactive, lode discoveries on Squaw Creek expanded the Pearl area's mineral potential northward across the Payette, primarily after 1880.

With silver mining ruined by price collapses in 1888 and particularly in 1892, interest in gold revived. Mines at Pearl, as a consequence, finally became productive in 1894. A production level of $30,000 in 1894 and again 1895 was increased to $80,000 in 1896. W.H. Dewey, who had become a wealthy Owyhee mine developer, took over the major mine at Pearl in November 1896. By the time his son, Edward H. Dewey, got through, a shaft had been sunk to a depth of 585 feet. Below the 400-to 500-foot level, sulfides became too much of a

Pearl in 1897

Lincoln mine at Pearl (c. 1900)

Pearl about 1904

problem, and operations had to shut down after about a decade of production. Another major property went into receivership, but was sold for $11,000 in November 1908. Extending the shaft to 540 feet in 1919 proved unrewarding, as did a limited amount of additional development in 1926 and 1932. Lesser properties had the same problem. Since Pearl's mines were not exactly worked out, interest in them continued long after significant production ceased. About 20,000 ounces of gold, valued at about $400,000 when mined, came from early operations at Pearl. Testing of old lodes at Pearl finally brought renewed activity to that old mining camp in 1980.

SOUTH MOUNTAIN

Two major veins of gold, silver, and lead were discovered in the fall of 1868 on South Mountain, not far from a new freight road from Silver City and Flint to Camp Three Forks and Camp McDermit. One of these veins was listed as from twenty to thirty feet wide on the surface; the other, although much narrower, still was four feet wide. Assays of $248.15 a ton gave South Mountain favorable notice right from the beginning. Development

did not get under way until after 1869, because a smelter was necessary to handle the ore. A small smelter was brought into the area in 1874, and until a financial collapse resulting from the failure of the Bank of California on August 26, 1875, shut down the district, South Mountain attracted a lot of attention. Interest revived in 1906, but production ($1.67 million) for the most part came between 1940 and 1945, with a lesser amount ($120,000) between 1950 and 1955. With encouraging silver prices in 1977, additional work was undertaken to develop ore so that South Mountain might be reopened.

SHOUP AND ULYSSES

Several independent prospectors based at Leesburg began to placer Salmon River bars at Shoup in 1868-1869, but lode discoveries there were delayed for more than a decade. Eventually Samuel James and Pat O'Hara came along late in 1881. Arriving at Pine Creek on November 24, they spent more than two weeks prospecting before identifying a major lode—the Grunter. Sam James later described their problems in hunting for gold to Jay A. Czizek (an Idaho state mine inspector who had lived at Shoup in 1889), who in turn reported their difficulties in detail:

> Undaunted by all obstacles which then beset their way, such as being the most unfavorable time of year for mountain travel, having to make their own trails in whatever direction they went, the river freezing and full of floating ice, but these were but trifling annoyances in comparison to a degraded remnant of a most fiendish tribe of Indians which then lurked about the hills.

Actually, Idaho's Indian wars had ended, and in any event, local Indians around Shoup were less savage than their white neighbors. But unaware of this, many prospectors had more than a passing fear, not totally unfounded, of an Indian menace.

John Ralston bought O'Hara's half interest for $250 before recording this important new claim in Salmon City on March 24, 1882. Then James sold his share for $5,000 in a much more profitable deal. James did not leave Shoup, though. He went on to discover one new lode after another. In this series of finds, he located the Kentuck (which became Shoup's other leading property) on June 17, 1882. That October he arranged to sell his new lode to Salt Lake City investors (including a former Utah governor, Eli H. Murray) for $16,500. H.C. Merritt managed to bring California's prominent mine developer, George Hearst, into this syndicate, which also included Gilmer and Salisbury, who operated a stage system in Idaho and Montana. A highly successful, professional prospector, James went right on discovering more lodes.

By the fall of 1882, Shoup had a post office (named for Lemhi County's leading citizen George L. Shoup, later to become Idaho's governor and United States senator, after the postal service rejected the local name of "Boulder" for the new community) and a bright future. The testing of ore commenced in November when a 6,300-pound lot, packed to Salmon for shipment to Salt Lake City, returned $375 in gold. Production in the Kentuck continued for many years after that without interruption—not too common a record in Idaho's mining history.

A substantial number of lesser lode locations followed the Kentuck. In 1886, Robert N. Bell (later to be another state mine inspector) joined Sam James as a successful prospector at Shoup, and George Hearst came in to make an additional discovery in 1888. By that time the town of Shoup had a couple of hotels in addition to the usual budget of saloons. Culture was represented by an opera house (which ran during the summer and fall) and an art gallery, operated by W.P. Pilliner, a professional artist. Not every Idaho mining camp could support such an array. In the fall of 1889, the Kentuck ten-

Shoup

stamp waterpower mill was grinding out more than $7,200 a month, and more than three hundred lodes—mostly low-grade—had been located. A Salmon River railway (projected by a Northern Pacific main line survey in 1872, but abandoned in favor of a route around Lake Pend Oreille) was needed to make more of the lodes productive. Most of Shoup's supplies came in by river freight boats from Salmon City, a service that had commenced when the Kentuck opened. Jay Czizek described what must have been an interesting inaugural trip:

> On the first day of December, 1882, five men and fourteen gallons of whisky embarked with a cargo of 7,000 pounds of supplies, which they safely landed on the tenth day out. Now two men will make the trip in two days with a cargo of 24,000 pounds, but whether the quantity of whisky is decreased in the same ratio as the number of men required, we are unable to state.

Shoup's original lode discovery had a less satisfactory development. Robert Bell later reported on that fiasco:

> The Grunter mine, situated a mile east of the Kentuck on the same vein, is a fine example of one of the most flattering gold enterprises in the State that was butchered by a would-be mining capitalist who, through blundering misconception of the enterprise he was undertaking, started in to put up a first-class ten-stamp concentrating mill for a half interest in the property, and wound up by furnishing a five-stamp mill with a hog-trough mortar and an overshot wheel that was just about as effective as a good sized coffee mill and never gave the property half a chance to show its merits.

Aside from Shoup, a small mining community two miles upstream on Pine Creek grew up after 1886 near another group of lodes. This settlement had not attained all the amenities available in Shoup by 1890. But life matched that in more than one isolated early Idaho mining camp:

A 10-stamp mill, a saw mill, one arastra and a boarding house are also company buildings. One half a mile further on is the burg of Pine creek, and two miles from here on upper Pine creek is the group of mines owned by James & Brown, and on this property is a one-stamp mill owned by Kenney & Pollard of Salmon City. The town of Pine creek includes the remaining buildings—which are built in a style of architecture similar of that of the raven, only differing in one respect, and that is where the raven has used mud to hold his structure together, mud has been used in this instance as a protection against the reverse elements of the weather. Crude and uncouth as these habitations may appear, they serve to keep up the illusion called home, but the romance of love in a cottage or in the cot on the mountain is often dispelled by the embarassing discomfortures of living in one with a family. We have overlooked the company dwelling house which is built of sawed logs and includes the company office and store house. This building bears the only outward semblance of a human abode, inasmuch as it has doors and windows. It also boasts the luxurious dimensions of two sleeping apartments and an attic which is very conveniently reached by a ladder put up to a window on the outside. The present appearance of these camps is by no means a disparaging feature, but marks an era of progress nearing the close of one decade, which has been attained without the spasmodic aid of a boom.

Shoup, for that matter, could not be compared with a more permanent community:

After eight years growth as a mining town, Shoup now contains two 10-stamp quartz mills, four arastras, one saw mill, one story two boarding houses, a postoffice and one saloon. No one has ever settled here with the intention of making a permanent home, therefore houses are an unknown quantity. The first glimpse of town, coming down the trail, is apt to give any one the impression of a collection of hen-coops, and though bearing the illustrious name of Shoup, the population of Chinese and Italians to be seen in passing through the town, suggests that it might more appropriately be called Pekin or Milan.

Mining went on for another decade or two, supported by large-scale lode operation. Mills with a total of 55 stamps had come into production by 1902, by which time over $750,000 in gold had been recovered. New discoveries not far away on Indian Creek increased the production of that area.

⚒   🐴   ⚒

Although gold lodes had been discovered in 1895 on Indian Creek, their location upstream from Salmon River delayed production until 1902. Access

was far more difficult than at Shoup, although Shoup's river boats sailed right by Indian Creek. Navigation problems came from lack of a good landing site. In 1899, when machinery was being taken there for a stamp mill at Ulysses, more than a little trouble was encountered. George M. Watson reported their rather interesting trip:

I left the mouth of the north fork of Salmon river in a boat on the 6th day of July for Indian Creek. The boat was loaded with machinery for a new stamp mill that is being built there.

Indian Creek is located down Salmon river, about forty miles from Salmon City. It is now creating great excitement since the recent rich strikes of free milling ores, and I think it will come to the front as one of the best producers of ores in Idaho. They are now building a 5-stamp mill there and are going to prospect the leads and if satisfactory will build two 10-stamp mills next summer. The ore is free milling, of a very high grade and is glittering with free gold, but after depth is reached the ores will likely turn into refractory or sulphuride ores the same as at Gibbonsville.

There is no road after you leave the mouth of the north fork of Salmon river, the only way to get transportation in there is by boats down Salmon river and that is very risky business, as the bottom of the river is strewn with machinery from recent wrecks.

Salmon river is a very bad river, as the writer knows by experience. Five of us left the mouth of the north fork of Salmon with a big flat-bottom boat loaded with machinery and supplies, and when we got to the mouth of Indian Creek we could not make the landing as the river was very high so we went down the river about a mile before we could stop the boat. Then we had to build a road back up the river to Indian Creek and drag the machinery after us, so you can see what kind of a country this is. It is insulated from every where and all it needs is roads and capital to make it one of the greatest countries in the west.

In spite of such obstacles, a fifteen-stamp mill, which commenced production in 1902, was enlarged to thirty stamps the next year. This operation processed enough low-grade ($10) ore to yield a monthly profit of three percent on a large capital investment through 1904. That September, a fire destroyed the plant and set Ulysses back seriously. At the time of the fire, Ulysses had Idaho's largest active gold mine. Even though production resumed, none of the mills ever managed to operate profitably in the long run. Their $600,000 production came at a loss, and since much of their low-grade rock could not cover the expenses of shipment to a smelter, Ulysses eventually proved to be a disappointment.

In contrast with Ulysses, early twentieth century mining at Shoup proved more successful. By 1908, Kentuck and Grunter development provided very large ore reserves, with some extremely high-grade pockets. Each mine had a successful ten-stamp water power mill, and two other mills (with twenty and five stamps) served nearby mines. Along with later production of $60,000 from 1935 to 1942, mining at Shoup and Ulysses finally accounted for more than $2 million in gold. Recent interest in molybdenum on upper Spring Creek continues to attract attention to the area around Shoup.

LOON CREEK

In 1869, placer mining excitements in Idaho seemed to be over. A series of departures of surplus prospectors, the latest being the exodus to White Pine in eastern Nevada, had relieved the territory of most of its idle mining population. Then the rush to Loon Creek reversed the trend. Although there had been earlier Loon Creek discoveries in 1864 and again in 1868, no one had shown any interest in them.

Nathan Smith, up to his usual prospecting, discovered paying Loon Creek placers. Taking out a party of Montana miners from Leesburg in May 1869, he found deposits that looked good; some pans went as high as $2. Then going on to Idaho City, his group obtained a saw and some supplies before returning to Loon Creek and then to Leesburg. Announcing his new find in Leesburg, Smith led a regular stampede of sixty to seventy fortune seekers back to Loon Creek on July 19. Upon reaching the new placers, Smith took up the ditch rights and the rest organized five districts, in each of which every miner was allowed to hold three claims. That way, the Leesburg group managed to control about fifteen miles along the stream for thirty days from the time that they began to take up claims on August 7. Some of the original claims were being located as late as August 25, and those could be held thirty days from then.

When news of the excitement reached Idaho City, the rush to Loon Creek then started from Boise Basin on August 14. As the newcomers arrived, there was a certain amount of discontent, because no more ground was available. The Leesburg owners were willing to sell but refused to allow prospective buyers to test the claims. Only two claims were thus sold, and they paid well. Once a claim had been held thirty days, it had to be worked or else it lapsed. But because the Leesburg locators could not begin to work all the claims they had taken, the monopoly on claims was expected to

be only temporary.

At the beginning of the rush from Idaho City to Loon Creek, packers and suppliers joined in, and there never was any serious shortage of supplies at Loon Creek. One packer, in fact, had taken a string of seventy mules loaded with merchandise out from Idaho City on August 14, and others soon followed. To those in Idaho City, Loon Creek appeared to be the biggest Idaho rush since 1864, as reported in the Idaho City newspaper of August 19:

> Miners, merchants, tradesmen, and excitement hunters have gone, and we learn that the exodus from Centerville, Placerville, Granite Creek, Pioneer, and Boston, is proportionately greater. Horses and mules, for riding and packing, have been in demand at high figures, and the excitement is as yet unabated.

Great excitement also seized Montana and a hopeful throng approached Loon Creek. Stories of large nuggets and of one pan running practically $18, with some others almost as good, kept up the Loon Creek fever. In August, some five hundred miners (by most estimates) reached Loon Creek, and soon the population rose to about a thousand. E.H. Angle, an old pioneer California miner, estimated 2,500 to 3,000 by late August. Almost no one on Loon Creek had a house, but the town of Oro Grande was being built, and the two hundred people or so who expected to remain during the winter were preparing for an unusually severe season.

The lack of lumber not only delayed the building of Oro Grande but also held back mining. Whipsawed at a high cost of $250 to $400 per thousand feet, lumber could be used only for essential sluices and mining equipment at first. Even so, claim opening did not proceed too quickly. D.B.

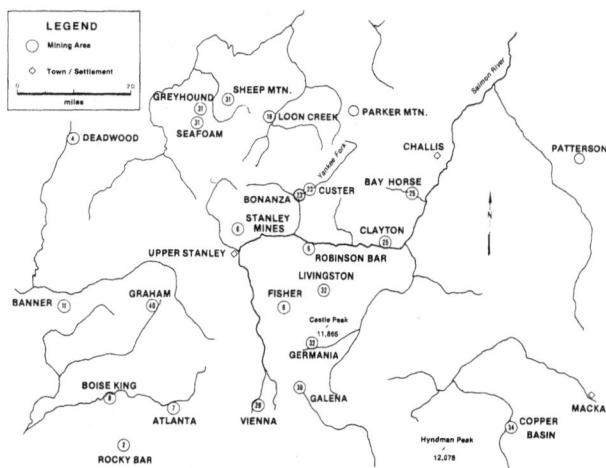

Loon Creek and central Idaho mining region. Numbers in circles refer to index map of mining areas, page 2.

Varney and company got started before anyone else. With five men on a sluice, they were recovering an ounce per day per man. By August 24, two more outfits were in production, and doing well. Varney's returns went up to $40 to $50 per man-day some of the time; precise figures for the first three days of September show successively $153, $170, and $120 for the group. Total Loon Creek production for the fall of 1869 was not expected to amount to much, but there were good reasons to anticipate a large return the next spring. Much of the stream could be prospected only with great difficulty, and some of the bars had six feet of boulders covering gold-bearing gravel next to the bedrock. Loon Creek gold, however, was regarded as about the handsomest dust in the northern mines, next only to Kootenai gold, if it was surpassed at all.

During the last week of August, quartz prospecting started at Loon Creek, when two veins—one assaying $31 a ton (ninety- four cents of this in silver)—were turned up. Aside from the lateness of the season, the camp was in a location extremely remote for quartz mining. For 1869, at least, nothing could be done to follow up these finds. But by September 22, twenty or more placer claims were open and producing.

Some degree of permanence came to Oro Grande in September. "Log cabins, canvas houses, and every other kind of habitable shelter of the mountain variety are going up every day," a correspondent reported on September 4, and a great number of real estate deals on town lots were in process. Twenty days later, Oro Grande had seven stores, seven saloons, one butcher shop, three boarding houses, two express offices, and ten to fifteen private dwellings. Thirty more buildings were under construction.

When the 1869 Loon Creek season came to an end, the mines were still paying well, and the next spring held great promise. By October 24, the weather was getting so cold that mining could not be continued very well, and all claims were laid over until July 15. Most of the miners from Loon Creek headed back toward Boise to spend the winter, and only about two hundred men remained in camp. When spring came, however, Loon Creek revived. Two hotels and a large number of new houses sheltered the permanent population. Two sawmills had produced 11,000 of board feet daily by the spring of 1870.

Even before the snow had melted over much of the trail from Idaho City to Loon Creek that spring, five different pack strings brought in enough supplies to take care of the needs of the permanent camp. Packing supplies through the snow belt section of the trail to Loon Creek had to be done

Loon Creek pack train

mostly at night: during the spring thaw, snow was too soft in the day for horses to get through. In spite of such obstacles, the population of Oro Grande rose to three hundred before the high water of spring runoff stopped travel all together. When the streams went down and mining could begin in earnest, all of Loon Creek was diverted into a long flume and used for placering a high bar near Oro Grande where most of the mining production was concentrated. Wages in that remote country were high for those days, $6 to $7 a day. But by July even Loon Creek had a surplus of labor. Some of the better placers paid well in 1869 and 1870, but others were regarded as suitable only for the Chinese, because the low-grade, poor-paying claims were rejected as worthless by white miners.

After the 1870 mining season, Oro Grande went through another winter with a population of about two hundred—this time including some sixty to seventy Chinese. Another season's mining really began to deplete the better placers and by the spring of 1872, Oro Grande's population of seventy-two was half Chinese. Only five (three whites and two Chinese) were women. Two express companies still served Loon Creek, although the indefatigable messenger for one of them was smart enough to work on his claim over on Yankee Fork until March, while the other company's expressman spent forty-six unfortunate days at the beginning of 1872 wandering in deep snow trying to reach Oro Grande from Idaho City. When the snow melted and the high water went down, some work still remained for white miners.

⚒ 🐴 ⚒

Soon after 1872, Loon Creek passed entirely into the hands of the Orientals. Six years later, on

February 12, 1879, the Chinese mining around Oro Grande came to a violent end. During the final winter, once proud Oro Grande housed only five Chinese. Located not far from popular winter camps of the Sheepeater Indians on the Middle Fork of the Salmon, Oro Grande had provided the local Indians with a convenient source of supply. White miners had also neglectfully abandoned equipment that the Sheepeaters found useful. The Chinese were less careless. They were also less able by their own few numbers to help provision the needy Indians during the long winters. The *Yankee Fork Herald* reported the confrontation as follows:

> [When] the Chinese were snugly in their warm cabins, with plenty of provisions on hand, Mr. Sheepeater made a call, and not meeting with that hospitality he thought due him on his own land, and his stomach calling loudly for that which he had not to give it, he resolved to do something desperate. After dark the Indians got together, and while most of the Chinese were sitting around a table in one of the largest cabins, engaged in the primitive and fascinating game of 'one-cent ante,' the Sheepeaters came down like a wolf on the fold, and the heathen, Oro Grande and all, were swept away as by a cyclone, while the victors returned to the bosom of their families on the Middle Fork to make glad the hearts of the little Sheepeaters with the spoils of the heathen.

The Sheepeaters later explained that they had no connection whatever with the disaster that befell the Chinese. But out of the Loon Creek-Chinese massacre came military expeditions that summer in which most of the Sheepeaters were rounded up — an affair known as the Sheepeater Campaign. After that, Chinese miners declined to work any more around Oro Grande.

More than twenty years went by before mining revived on Loon Creek. Then, the 1902 gold rush to Thunder Mountain about thirty miles north and slightly to the west brought in prospectors by the thousands. Rich float found in early days, "accounts of which remained a folklore of the region for years afterwards," at last was traced to a vein which lacked prominence on the surface but which soon showed great promise. This lode — the Lost Packer, discovered by Clarence Eddy in July 1902 — led to another round of excitement. (Eddy was particularly gratified at his luck, since he at last could afford to publish a book of his poems and to start a newspaper in nearby Thunder Mountain and Custer.) Mules packed out five carloads of rich ore in the summer of 1903, and several more in 1904. This kind of cumbersome shipment was subordinated to road construction in 1904, so that the two-year job of hauling in a modern smelter (with a capacity of 100 tons) could be completed the next

Lost Packer smelter

summer. An accident at the start of operations in 1906 delayed production another year, but the smelter ran profitably for thirty-four days in 1907.

By 1914 about $500,000 production of copper, gold, and silver had been realized. The Loon Creek road, moreover, enabled another company to rework the old placers with methods less crude than operations of the early days. The high cost of mining in a district so difficult to access limited production greatly; lodes such as the Lost Packer simply could not be worked on a large scale during the short season that the high mountain road to Loon Creek could be kept open. Only the richest ore could be processed, and much of the original expectation for Loon Creek never materialized. Exploration of some Lost Packer veins in 1980 brought a prospect for revival of that property before price declines set back silver mining again. This is one of a substantial number of Idaho mining properties which a more advanced twentieth century technology might be able to revive.

YELLOW JACKET

During the fall of 1869, almost everyone at Loon Creek lost interest in building up that nevertheless promising new mining camp. Nathan Smith, back from another prospecting tour, had another startling discovery to announce — this time, the Yellow Jacket.

His Yellow Jacket party thought that they had another Boise Basin, and a stampede on September 24 to the new bonanza depopulated Loon Creek. Some four hundred men took off with Nathan Smith, only to find the new district vastly overrated. In the words of John Ward: "Gold is very scarce in Yellowjacket, but the broken down horses and mules are plentiful along the road." The trouble had been that one of the members of Smith's Yellow Jacket discovery party had heavily salted the prospectors' pans, apparently with California gold, and then had thoughtfully disappeared before the rush to Yellow Jacket revealed his deceit. Smith was as disgusted as everyone else at being the victim of a practical joke, and by the time the stampeders had all got back to Oro Grande, there was "terrible swearing on Loon Creek." An incidental result of the hoax was the immediate discovery of some important Yellow Jacket quartz leads that eventually proved to be productive.

In 1876, a prospector's three-stamp mill was completed to test the district, and six years later, arrangements were made to import a larger plant. In April 1883, packers loaded a ten-stamp mill onto mules and dug through snow drifts up to twelve feet deep in order to get into operation by June 1. That way their water driven mill did not miss a season when power was available. Ore was freighted with two wagons (each with four horses) down a mile and a half grade to the mill site at a cost of $2.50 a ton. Sleighs were used in winter. Like the stamp mill, which processed thirty tons of ore a day, both wagons and sleighs had to be packed into Yellow Jacket. About thirty miners were employed until October 1892. Then a $100,000 mine purchase by Colorado investors led to a major expansion of activity there.

After a two-week $3,000 cleanup late in 1892, Yellow Jacket's new Colorado owners saw that they needed to invest in a more economical production system. To reduce the costs of getting ore to their mill, as G.L. Sheldon explained it, they

> decided to erect a Swem aerial tramway, the buckets to carry 125 lb. of ore each. No packer would contract to deliver the ⅞-in. wire cable required in its construction. The company's pack train brought in the cable, 8,400 ft. in length in three trips. Being too stiff to coil for individual coils on each mule, it was strung out upon the main street of Challis, six or seven runs on a side being tied together. The mules were placed in the center, with the cables lashed to each side, the loop at either end swinging clear of the leading and the end mules. Nearly all the inhabitants of the county were on hand to see the pack train start. They had plenty of excitement and fun. It took two men to manage each mule for the first few days. On uneven ground the individual loads would vary in weight. In a hollow the rope would lift the center mule off its feet. On a ridge or knoll one mule took the load of three. One wall-eyed cuss bucked and tore around on a ridge, throwing the whole pack train of twenty-mules down the mountain 150 ft. into the timber in a tangled, twisted condition. It took two days

to cut them out, no serious damage being done.

> Owing to the stiffness of the several cables bound together the pack train could not make short turns, and a temporary straight trail, regardless of grades, was therefore made. Eventually the mules became accustomed to the novel loading, and the entire cable was delivered without serious mishap. The tramway reduced the transportation cost for delivery of the ore from mine to mill to seven cents per ton.

During the Panic of 1893, Yellow Jacket's superintendent, fearful that his miners would be unpaid, refused to ship the June bullion production to Salt Lake. That action almost resulted in forfeiture of the mine. But G.L. Sheldon found out what had happened, sent the superintendent out prospecting for a new mine, and made the payment barely in time to avoid delinquency. Sheldon then took over and managed Yellow Jacket's major property for two years. He still had to overcome problems arising from his isolated location. The replacement of a worn-out 625-pound camshaft proved difficult, but Boise's noted Basque packer, Jesus Urquides, could handle heavy loads:

> He secured the largest mule in the locality. He then made two tripods the height of the shaft when loaded. These were packed on another mule. The big mule was led with the load, one, two or three hours, depending upon the condition of the trail. [Urquides] would then stop and set up the tripods just behind the loaded mule. Four men would next slide the shaft back onto the tripods. The mule was then allowed to rest and feed for a short time and the procedure repeated.
>
> The mill was operated by a Leffel water wheel, which was connected to a penstock forty-two feet in length. A ditch, about 1,500 ft. long, conveyed the water from a six-foot dam on Yellow Jacket Creek. Anchor ice that formed on very cold nights in the creek would sometimes, when the temperature rose suddenly, break loose and in the form of a fine, slushy material, before we know it, fill the penstock full of ice, stopping the water wheel. It took two days to dig it out.
>
> The ditch on the hillside would often break, causing many shutdowns.

By making other improvements as well, Sheldon kept his Yellow Jacket mill in operation until mid-May 1894, when he lost his entire plant in a fire.

While Sheldon was reconstructing his mill, C.L. Coleman — a major Wood River investor who had to shut down there after silver prices collapsed in 1888 — turned his attention from Hailey, Vienna, and Sawtooth City to Yellow Jacket. Gold mining proved attractive during hard times in the summer of 1894, and two hundred men were at work there. Coleman spent $50,000 putting in a large flume for

Yellow Jacket mill (photograph taken in 1981)

his placers, and two new stamp mills were under construction. By salvaging and rehabilitating ten stamps from his burned out twenty-stamp mill, Sheldon was able to enlarge his plant to thirty stamps when a twenty-stamp replacement arrived that fall. Sheldon also got a sawmill to provide a half-million board feet of lumber that summer so that he could build a 75-by-150-foot structure to house the new installation. Plans for his plant reached Yellow Jacket, October 4, 1894, and he got his mill packed in there in a record 106 days that fall and winter.

Sheldon's new plant proved highly efficient. With a milling cost of only $2.67 a ton, a monthly yield of $50,000 was realized from free-milling ore. A six-foot Pelton wheel under 150 feet of pressure generated electric power to operate Yellow Jacket's thoroughly modern cyanide mill which had a 200-ton daily capacity. Designed to serve a mill with 500,000 tons of ore (estimated at $7,500,000 in total value), this plant was sold in May 1895 after several months of successful operation. Under new ownership and management, it was enlarged from thirty to sixty stamps (contrary to G.L. Sheldon's advice) at a cost of $72,000. But only $68,000 was recovered from that investment, because the $7 to $10 low-grade vein soon was lost in a fault.

Yellow Jacket continued to progress for most of a year before a general economic collapse discouraged most miners. On September 1, a wagon road, costing $4,500, was completed from Challis and helped reduce Yellow Jacket's isolation. But just then a smaller Cleveland company went broke, leaving its twenty employees unable to collect their back wages. With 175 miners remaining at work, Yellow Jacket entered 1896 with bright prospects regardless of that failure. At the beginning of May, a million dollar purchase in New York brought in new management and new troubles. A dispute over

a nine- or ten-hour work day, optional use of contracting rates instead of hourly pay, and employment of non-union miners led to a strike that shut down operations later that month. Reopened July 8, 1896, after settlement of these difficulties, Yellow Jacket ran through September before closing again because of failure of the milling system. Effective for free-milling ore that it had been designed to process, Yellow Jacket's large mill could not handle values available after that ore supply unexpectedly ran out. Considerable exploratory tunneling failed to disclose any more ore, and Yellow Jacket's handsome sixty-stamp mill remained idle from then on.

Many of Yellow Jacket's miners — Populists almost without exception — interpreted this operational suspension as an effort to drive them out of camp prior to election day in the fall so that they could not vote. (Such tactics were employed in that exciting election in the East. But Idaho's silver forces had overwhelming strength. They could render any effort to discourage Populist voters futile, by closing mines or by any similar device.) With only fifteen out of seventy-five to a hundred miners employed, Yellow Jacket sank into a considerable depression. Three saloons, two stores, and two restaurants provided more facilities than a camp needed when most miners were headed to Gibbonsville and other mines.

⚒   🐎   ⚒

Reduced operations at Yellow Jacket coincided with encouraging activity nearby on Silver Creek. Discovered in the summer of 1876, an extension of the Yellow Jacket lode held great promise. After further testing in Denver, several recovery processes were evaluated for efficiency in handling new Yellow Jacket ores from Silver Creek. Then a 300,000-pound Clareci-Pellitan cyanide plant with a roller crusher device was freighted in. Elegant cyanide plants, however, failed to justify their substantial expense and effort in installation. Conversion costs to cyanide, along with burdensome initial investments and development charges, appear to have exhausted the financial resources of Yellow Jacket's primary company, which defaulted on a $172,792.80 obligation to John T. McChesney of New York. McChesney acquired the property for $172,000 at a sheriff's sale in Salmon on August 16, 1897. In effect, he acquired more than a million-dollar value for his unpaid loan. Yet he managed to operate his new property for only a month before having to shut down again in order to obtain an effective recovery process. With fine modern conveniences, including two trams, two Pelton

Yellow Jacket bunkhouse (photograph taken in 1952)

wheel electric generating systems (removed from Idaho Falls), and other technological advantages, Yellow Jacket still lacked an element essential for production. In 1901, John E. Searles, who added Yellow Jacket to his large financial empire, went bankrupt in May. In 1908, the unfortunate camp still had two large modern milling plants that had been used only slightly.

With eleven hundred feet of development tunnels and a very large volume of potential ore averaging around $1 a ton, Yellow Jacket attracted enough attention in New York so that a new group of investors decided in 1909 to find a technological solution to the milling and recovery problems there. Reduced freight costs encouraged considerable activity in 1910, and Yellow Jacket's large modern cyanide plant was ready to operate as soon as the rather complex ores could be treated successfully. Development work continued through 1912, when consideration was given to reworking tailings of $12 ore which had been milled in 1896 with considerable loss of mineral values. After efforts to discover an effective technology for Yellow Jacket ores failed in 1912, a small placer recovery came in 1914. A revival in 1922 produced a small ore shipment that encouraged a Salt Lake company to build a five-story hotel. Funds ran out January 15, 1924, prior to completion of the large frame structure, and fifty-two years later its roof caved in from winter snow.

With financial failure preventing lode mining, another disappointing effort to operate a hydraulic giant followed in 1924. Development work finally resumed in 1929, with still more low-grade ore exposed. A few tons of selected high-grade ore were shipped out then, and several old properties were expanded. Reduced costs of mining during a nationally depressed economy allowed additonal development of several important Yellow Jacket properties by 1932. About eight miners worked

there from 1929-1932, driving several hundred feet of additional tunnel. One company attained more than two thousand feet, and others expanded their tunnels also. By 1936, a major development tunnel had been driven 2,700 feet, and others were deepened also. A few carloads of ore were finally shipped for smelting at Anaconda, Montana, and at Garfield, Utah, west of Salt Lake City. These tests failed to offer profitable returns: an investment of about $40,000 produced a $10,000 yield. (Concentrates shipped to those smelters recovered close to production costs, but ran far short of returning the costs in developing 430 feet of tunnel.) So a sale was arranged to Salt Lake investors in 1938. By that time, several miners had been employed there for about a decade, and their main development tunnel reached 5,000 feet. But from more than a mile of underground workings, no ore was being produced with available recovery technology. Yellow Jacket shut down again for fourteen years.

Eventually a flotation mill was started in 1953 to serve a property almost a mile from Yellow Jacket, but that revival did not last very long. Three miners went to work again in 1969-1970, but Yellow Jacket ores still could not be treated economically. With a large body of low-grade ore located at a great distance from a smelter, and with difficult recovery problems to compound Yellow Jacket's disadvantages, that camp retains a spectacular 60-stamp mill that never has had the opportunity to get worn out. An old, established Yellow Jacket mining family still is ready to resume operations there in 1982, once difficulties of production are overcome.

Yellow Jacket's mineral production may have approached $400,000 but most likely did not come too close to that amount. Largely from four and a half year's work, that total included scattered operations ranging over more than a century. Five mill runs, mainly in 1895 and 1896, contributed $121,761.56 of that yield:

November 2-December 10, 1893	$4,060.03
February 3-June 26, 1894	$12,086.20
March 1-December 26, 1895	$60,991.11
January 6-December 15, 1896	$42,050.76
June 3-11, 1897	$2,573.46

These results from operating one of Idaho's largest stamp mills for several years (with a 92 percent recovery reported from low-grade ore) indicate the exceptional difficulty of mining there. With an average of only $5.50 per ton, even Yellow Jacket's best production was marginal. After that, complex ores could not be processed satisfactorily. As a result, Yellow Jacket's challenge to mining engineering still has not been met.

FROM GOLD RUSHES TO GOUGING: MINING LESSONS LEARNED BY 1869

Compared with early placer operations, quartz development had scarcely started by the end of the first decade of mining in Idaho. Most lode production by 1869 had come from Owyhee. Stamp milling in Boise Basin, South Boise, and Atlanta had faltered in its early beginnings, and although confidence was great that milling soon would pay, the day of really successful quartz mining was farther off than operators then generally supposed. Placer mining, with its simpler technology, had posed much less of a problem. Although some major areas, particularly Boise Basin, remained active for many years and continued to yield large returns through later dredging, the future of Idaho's economic development after 1869 lay largely with lode mining.

Even if stamp milling had succeeded in the various Idaho camps where it was tried initially, lode operations would have lasted for a relatively long time, and the mining communities, along with those service communities which supplied them, would have prospered over the years on a basis much more permanent than ordinarily could have been expected from placer mining. Mining failures, it turned out, gave the mining and service communities an even longer future. The inability of the early companies to operate successfully meant that the mines lasted many extra years. Part of the capital investment of the unsuccessful companies went into building the communities in which they operated. Mining failures as well as mining successes contributed to the permanence of the economy of the general mining region.

The reasons for early stamp-milling failures varied somewhat with the company and with the mining district, but some were repeated quite widely. All of the Idaho districts suffered severely from lack of good rail transportation; some were more remote and more inaccessible than others, but none of them had really successful mining until after 1882, when rail service began to reach the Idaho mines. Improvements in technology were also needed. The introduction of dynamite helped considerably in all the districts. Improved devices for pumping water, for example, were unavailable until later; South Boise, for instance, was held back for years because the major producer there was under a large stream which had to be pumped out. Expensive title litigation often ruined or retarded many of the earlier, better properties. The inability or failure to consolidate claims into mines of reasonable operating size compromised many ventures, especially in Owyhee. The lack of dependable, experienced labor held back the more remote districts especially, but as the population of Idaho increased, labor became less of a problem.

Experience in mine administration was also a necessity for successful superintendents, and as time went on, better qualified management developed. Not very many of the first managers blocked out ore in advance of bringing in a mill. More than one enterprise had failed when it turned out that its mine simply lacked the ore essential for operation. Others failed, or were retarded seriously, upon finding that the wrong kind of milling process had been installed. Practically all of the districts could have done much better with fewer reckless or dishonest mangers who, in all too many operations, neglected to pay labor, equipment, and transportation costs and who, in some instances anyway, simply absconded with the proceeds of the mine. Dishonest promotions were a menace that did not disappear as time went by; investors, though, eventually became a little more alert to this hazard.

For the honest enterprise, the failure to anticipate the amount of capital needed to get a mine open and a mill in operation often caused trouble. Stockholders expected, without putting up the cost of a mill, to have the mine and mill return dividends without the essential preliminaries of getting the mill into the district, getting the ore out of the mine (assuming some ore was there, a consideration which usually was left to chance), and then getting the milling process adapted within a trial run or two. Almost invariably, the company would have to raise capital beyond the amount originally planned, whereupon the distant stockholders, unable to tell for themselves what was going on, often suspected fraud and let the company fail rather than risk further loss. In too many ventures, inferior mills that never would work satisfactorily were brought in at great effort and expense, with results indistinguishable from outright fraud.

In assessing the causes of early failures in lode mining, allowances must be made for the special difficulties which confronted early operators. Inexperienced in evaluating mineral deposits in new country, developers had little reason at first to foresee that rich surface prospects would not hold up at depth. Before dynamite replaced black powder in 1868, the costs in drilling the tunnels and shafts essential to determine if available ore justified bringing in an expensive mill exceeded whatever resources most developers could command. Miners in isolated districts often faced an extra year's delay if prudent development preceded their investment in mills. They simply could not afford to wait long enough to employ such a cautious approach. Investors normally would not tolerate such delay. Even if they had capital resources adequate to permit them to mine a large ore reserve prior to bringing in a mill, they would have faced severe problems. High graders would have appropriated the best ore and pounded it up for a profitable return. The installation of ore bins of a size sufficient to retain reserves adequate to amortize a stamp mill would have posed great difficulty at many locations. For a new mining region, any effort to employ an approach, which later experience justified, would have been difficult even if managers had thought to try. If they had resorted to development practices later identified as appropriate, many of their stamp mills would have failed to overcome technological recovery problems — particularly those associated with sulfide ores. Like farming and many other economic pursuits of the time, lode mining presented a complex series of hazards to investors and employees alike.

Considerable attention was devoted at the time to analyzing a variety of reasons for the early failure of stamp-milling enterprises. But a spirit of optimism generally prevailed. The founders of mining communities felt that they were establishing stable, permanent settlements, not potential ghost towns. Their efforts were directed largely against the frauds, on the theory that the mines, given an honest development, were good and could not help but pay. When, for example, Dr. S.B. Farnham, in explaining his South Boise failure, pointed out some of the difficulties of operating with inadequate New York capital, the *Owyhee Avalanche* rejoined with the devastating exposure of the problem created by the frauds:

> The discovery of rich quartz . . . created immense excitement and opened new channels for business; it soon became apparent to the shrewd ones, that a system of humbugism could be inaugurated to the benefit of many branches of trade; and that all the capital necessary was the muscle and courage of the hardy prospector. The consequence was that whenever a silver- or gold-bearing lode was discovered, a thousand claims were located by these hums, claiming to be upon the same lode or upon others far better and richer, and as there are many deposits of rich ore in the country, it was not difficult to show specimens of the precious metals, and far easier for these men to represent that they had come from the location in question. Now as paying mines in any country are known to be worth many thousand dollars, the only question with these hums was whether they could convince capitalists that their interests had merit; this difficulty was very easily overcome, for Prof's A., B., and C. will make a 'Geological survey, full and scientific report to order, for a portion of the proceeds of the sale; assayors will make the proper returns for the same price; mining engineers will make a survey and plat, showing the Topography of the country, timber, mill site, & for shares; Judges, Generals, Majors, Captains, Colonels, Bankers, and Merchants will endorse anything for the asking, and a small portion of the proceeds; the Recorder will give abstracts of title, embellished with any quantity of red tape and sealing wax, for pay; the thing is complete.' Mr. H. then arms himself with a box of specimens — twice out of three times procured from the dump of some paying mine in the country; — thus equipped he goes to the East to impose upon men, who, as I said before, are not used to seeing Professors, Judges, Governors, Bankers, and Merchants endorse men of this character.

No one should be deceived into thinking that this kind of operation characterized only mining promotions. During the Gilded Age after the Civil War, mining swindlers had a hard time keeping up with all of the other frauds across the country. The scandals of Reconstruction in the South and of the Grant administration nationally were only another sample of a deluge of dishonesty that swept the nation during and after the Civil War. Examples could be found equally well in the petroleum boom in the East that matched the gold rushes of the West in surprising detail; in the construction of transcontinental railways as well as a host of other railways; in the industrial expansion of every variety, particularly in the Northeast; or in such special manipulations as Jay Gould's celebrated attempt to get control of the U.S. Treasury's gold reserve that brought about the spectacular Black Friday Panic of September 24, 1869. By comparison, when it came to deceit, the promoters of Idaho's mines were pikers.

For about two decades after the initial stamp-milling failures, lode mining continued with only partial success at best until large-scale operations finally became practical. Much of the work was decried at the time as "gouging": Operators interested only in short-run profits would work selected high-grade portions of their veins, disregarding the best general development of their mines. Much of the early rich Owyhee production may have been gouging, and laments over that practice began to arise from other districts when milling attempts resumed after the initial failures. As one manager succeeded another, or as often was the case, as one gouger succeeded another, the new superintendent more than likely could not learn what the old one had done. The Poorman in Owyhee was plagued especially by this kind of discontinuity in management. In exceptionally difficult cases, the old staff would not tell the new management what had been tried in the past, or what processes had worked or failed. A new superintendent might bring in his staff and try different methods, entirely unaware of previous experiments for whatever that knowledge might have been worth. In some instances, superintendents could not even manage to ascertain the mine's previous development and what had been learned. Few situations, naturally, were as extreme as some of these. But during the age of the gougers, continuity in mine development was deplorably lacking. Lode mining somehow continued during those clumsy years and bridged the gap between the early placers, which had laid Idaho's economic foundation, and the large-scale, base-metal mining that along with successful gold and silver quartz mining gave Idaho the economic boom necessary to bring about statehood in 1890.

Part II. New Camps and Chinese Mining, 1869-1878

INTRODUCTION

Gold discoveries in southern Idaho continued to encourage prospectors and miners after the rush to Loon creek in 1869. Yankee Fork (an offshoot of Loon Creek), Cariboo Mountain (south of the Snake River not far from Wyoming), and Black Pine (close to Utah and the Kelton road) followed in 1870, and the Bay Horse was prospected successfully in 1872. Important properties along Yankee Fork eluded prospectors until 1876; the Yankee Fork gold rush came mainly in 1878 and 1879. Excitement over Cariboo Mountain, eventually responsible for 60,000 ounces of gold, failed to match Idaho's more dramatic earlier mineral discoveries, and Black Pine lacked enough ore to create any major commotion. By 1869, most of Idaho's larger gold districts had been prospected to some extent. Yet much remained to be discovered, and mining aside from gold and silver had to wait for another era of mineral development.

Conventional prospecting methods, adequate to locate gold claims in places such as Boise Basin, did not disclose many of Idaho's better mineral possibilities. In Idaho, as in many of the world's most impressive mining regions, professional prospectors supposedly knew better than to investigate some places that nevertheless eventually proved more than worthy of attention. (Gold prospectors in the Yukon, for example, overlooked multi-million dollar opportunities on the Klondike for quite some time while checking out that future mining area. In South Africa, the world's major gold resources appeared in a formation that no competent prospector considered had any reasonable chance for an extensive deposit.)

Idaho similarly had lodes that experienced prospectors regarded as not worth looking at. As a result, gold seekers who had little or no idea of what they were doing sometimes came by accident upon mineral possibilities that informed prospectors would not have noticed. Through various approaches, in which prospectors gained the experience that helped them anticipate new mining possibilities, or in which the massive inspection of new areas by novices occasionally paid off handsomely, new mineral opportunities were identified.

Mining discoveries after 1869 often followed a course somewhat different from Idaho's earlier gold rush pattern. Placers and gold lodes could be prospected far easier than copper or lead-silver outcrops could be found. In general, silver tended to follow gold, and lead-silver, zinc, and copper came still later in this sequence. Gold mining often led to still larger production of other metals. Nothing in southern Idaho compared with substantial Coeur d'Alene lead-silver discoveries farther north that were delayed until 1884, when a gold rush brought in fortune hunters who turned up lodes that far eclipsed anything Idaho had to offer previously. Times changed substantially after the gold rush era gave way to another, more subdued period of mineral development which eventually led to dramatic mining possibilities in the Wood River country in 1880 and in Coeur d'Alene country after 1884.

By 1870, Idaho had more Chinese miners than any other state or territory, and the majority of Idaho's miners were Chinese. Oriental miners had little choice but work lower paying claims, for they usually were admitted only to the poorer areas remaining after major gold rushes had cleaned out those highly rewarding richer placers.

A number of new lode camps found after 1869, while unsuitable for Chinese operations, started with inferior prospects reminiscent of Chinese placers. Most of Idaho's newly discovered mining areas for a decade after 1869 got off to a slower

Chinese miners operating a giant at Rocky Bar

start. By that time extensive prospecting had disclosed many, but by no means all, of Idaho's potential gold camps. Some of those which followed offered more opportunities for Chinese miners than for whites.

Chinese operations fitted in well with Idaho's newer mining trends. Snake River fine gold properties, located in a number of places in 1869, characterized a new era in Idaho mining. Early Yankee Fork operations also started slowly before building up to a substantial gold rush toward the end of a decade of modest development. Cariboo Mountain, Black Pine, Bay Horse, and Gibbonsville went through similar slow development phases representative of the hard times accompanying the Panic of 1873. With national economic recovery after 1878, Idaho mining developed along more prosperous lines. A surge of interest in lead-silver properties, largely dormant prior to 1880, helped stimulate mining along new lines that developed new districts quite different from most of Idaho's earlier operations.

During Idaho's second decade of mining, old established camps went through a similar era of recession. Aside from a conspicuous exception at Silver City until 1875, placer and lode properties generally showed less opportunity for fabulous profit than had been anticipated a few years earlier. The lack of transportation, combined with problems in management, promotion, and technology, held back lode mining in existing camps as well as in new ones. Many mining areas discovered after 1869 illustrate problems applicable to most of Idaho's mineral development of that era.

SNAKE RIVER

Gold mining in the Snake River preceded the more celebrated discoveries at Pierce in 1860 and in

Boise Basin in 1862 by a number of years, but was conducted on such a modest scale as to attract almost no attention whatever. As an indirect consequence of the Ward massacre of 1854, placer mining at Fort Boise (located adjacent to the junction of Boise River with the Snake River) had provided recreation for men of the military force under Major Granville O. Haller sent in the spring of 1855 in reprisal against the Boise Shoshoni Indians. While camped at the fort, soldiers noticed interesting placer ground in the immediate vicinity. (A generation later a dredge operated on the Idaho side of the river right next to the fort site.) In the six weeks they were there, these men mined a pretty display of gold. But the deposit they were working on certainly was not commercial by their methods, and no one followed up their limited discovery. This episode passed unnoticed until John H. Scranton, who had seen the product of the early Fort Boise placers, remembered the find a few years later and recorded it in Lewiston's *Golden Age* (for which he was editor) after the rush to the Boise mines got under way in 1862 and had attracted attention to the locality.

Interest in the possibility of mining on the Snake River revived with the Boise Basin excitement, and an unfortunate rush of some two thousand miners to the upper Snake late in the summer of 1863 led to a great disappointment that discouraged prospecting of the river. An account of prospecting problems appeared in the *Boise News* (Idaho City), October 20:

> That there are good mines on the head waters of Snake, is altogether likely. Yet we can see no positive evidence that any thing has been struck to warrant the mighty rush of people that are now on their way pursuing — as far as their own knowledge of the country, or the existence of any mines in that direction — a phantom, a mere myth. If it should turn out that there are good placer diggings out there, it will all be well enough, and their trip with its hardships and privations will not have been in vain, but if — as is most apt to be the case — it should prove otherwise, there will be no such thing as estimating the amount of suffering that will be undergone by those of small means and no preparation for Winter. We have been informed by a party who has returned from a fruitless search in that country, that those who have reached the Blackfoot fork of Snake are utterly confused and have no other goldometer to guide them than horsetracks, every one of which is taken as a sure indication of the existence of gold, and that the Snake river mines are in the direction to which they point, but after following them until they run out or turn back, they have thus far had to retrace their steps and seek similar indications in

other directions. Some have abandoned the hunt and gone in quest of fortune in the Stinking water region, while others persist in a determination to find it on the Snake. That gold does exist there we have no doubt and that it is coarse and heavy, is almost equally certain; at any rate we have been shown some very large pieces and assured that they came from the base of the Wind river mountains, and have no reason to dispute the fact. The mines may be of that character and be at the same time extensive and good, yet in our mining experience, we have observed that coarse gold is not always the best to mine for — it is apt to be spotted, and a few may make a fortune while the masses almost starve.

Yet some Snake River placers, like a number of Salmon River bars, employed limited parties of miners near Lewiston far to the north in 1863 and 1864. Efforts to expand Snake River operations farther upstream continued also. On July 15, 1870, a correspondent informed the *Daily Evening Bulletin* (San Francisco):

> In the summer of 1864, a well-appointed company left Boise to prospect the upper Snake and its tributaries; but, meeting with resistance from the Indians, they were obliged to abandon this purpose, having only ascertained that there was at least a show of gold all along the streams in that region. Ever since these attempts have been annually renewed, only to end for the first two or three years in similar results.

Aside from the difficulty in finding bars rich enough to return a profit, Indian resistance continued to discourage efforts to search the Snake until General George Crook managed to bring the Snake war to a halt June 30, 1868. Then in 1869, a severe regional water shortage left a host of unemployed miners who found prospecting along the Snake less expensive than paying room rent in Boise or in Idaho City. Many who did not go to Loon Creek examined many Idaho locations. The Snake River offered a superior opportunity that season because low water levels facilitated channel prospecting which could not be undertaken during a normal runoff.

The discovery of workable placers on the Snake River in the neighborhood of Shoshone Falls resulted from careful and systematic prospecting by an old associate of E.D. Pierce — a man named Jamison. "Being satisfied that gold existed on Snake River," in the summer of 1869, Jamison started "up its banks from near its mouth on a general prospecting tour, but did not find gold in sufficient quantity to work until he reached the vicinity of the three islands on the Bruneau River." He did not regard anything as really profitable until he got to Shoshone Falls. There he had a good base for operation on a major new road. Construction of the

Central Pacific railroad had led John Hailey to commence a stage line over a new route from Boise to Kelton in June 1869, and his coaches crossed the Snake River at Clark's Ferry near these new Shoshone Falls placers.

For a time, Jamison's company worked in a rich eddy about three hundred yards above Shoshone Falls, where it was possible to make $40 a day for awhile. Then they moved six miles below the falls to a point near the mouth of Rock Creek, where they continued to work early in the spring of 1870. By February, reports of Jamison's discoveries led quite a few prominent prospectors to look at the Snake River; they found, though, that there was not much to get excited about. Three men were making $8 a day at Shoshone Falls with a rocker, if they were not exaggerating. Even by mid-March, there were altogether only four companies (a total of twelve miners) with five rockers making average wages in the entire Snake River placer operations. Some of

those who went out to examine the new mines decided that the main reward for their trip to Clark's Ferry was the spectacular view they got of Shoshone Falls, which they regarded honestly as making the whole effort worthwhile. The recovery of Snake River fine gold scarcely had begun at all in the discovery year of 1869, but in 1870, new placers were found in many stretches of the Snake.

By 1870 although the majority of Idaho miners were Chinese, they did not meet a good reception near Shoshone Falls. A white miners' convention on May 18, 1870, resolved to exclude Orientals from their camp. As in Boise Basin by 1870, whenever Chinese miners invaded a district, placers remaining there were already regarded as too low grade to interest whites. Chinese miners willingly worked these poorer claims, primarily because they earned far more than they would have made in China. Even though they were usually denied by the white mining districts an opportunity to profit

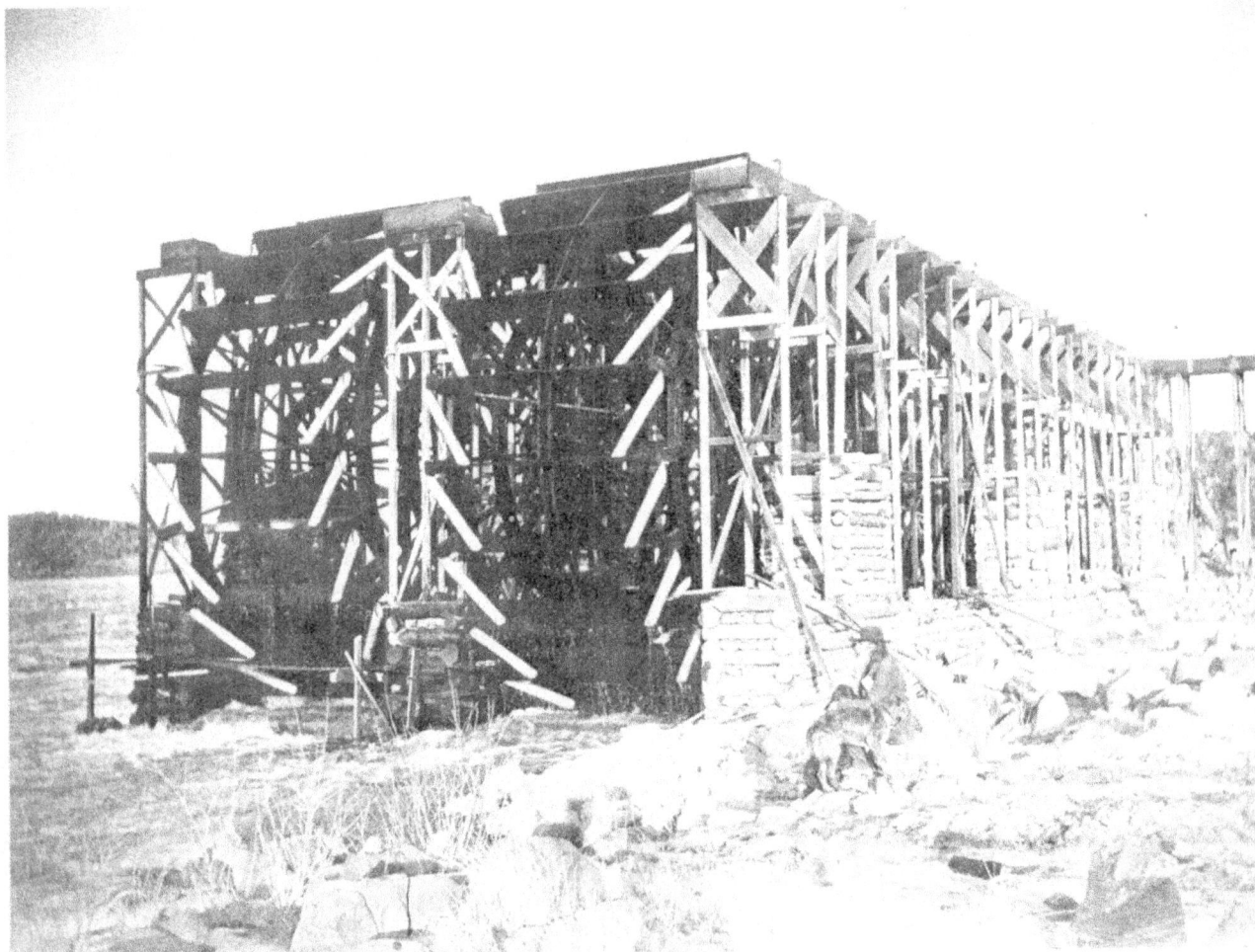

Waterwheels above Milner supplied a flume for sluicing fine gold

An elaborate fine gold recovery process above Milner after 1880

from any of the better mines in the West, the Chinese planned to retire to their homeland with enough wealth to improve their lot substantially. By 1873 the Oriental miners got a public welcome to return, for the economic reason that the white miners could find no one else who would buy these inferior claims.

As in 1869, the Snake River ran at a low level in 1870. In normal years, high water made prospecting difficult or impossible until the middle of August. So in 1870, when low water again forced large numbers of miners to leave their regular jobs, the Snake could be examined again. On July 15, a *San Francisco Bulletin* correspondent reported that extensive prospecting had shown considerable promise:

> Through these persistent [prospecting] efforts, prosecuted both from the east and the west, the main stream has been traced and examined quite to its source, in the Wind River Mountains, while most of the upper tributaries have also been pretty effectually explored. On nearly all the bars, both on the two principal forks, as well as the confluents, gold has been found—always excessively fine and generally only in limited quantities—nowhere in very great quantities.

In the Shoshone Falls area, only a few elevated bars could be prospected or worked successively during high waters. So facilities were limited. A camp at Dry Creek (just below Caldron Linn near later Murtaugh) had four stores, a restaurant, and about six residential tents. Shoshone Falls had a store. At Twin Falls (on the river above Shoshone Falls), "Shoshone City, the largest hamlet on the river, consists of four canvas shanties and a tent, all used as trading posts."

In the spring of 1870, enough prospectors had swarmed up and down the Snake River to locate workable placer ground in many scattered places. Aside from Shoshone Falls, miners found enough gold to justify permanent camps near J. Matt

Taylor's bridge (at the site of Idaho Falls) as well as around Salmon Falls in Hagerman Valley. Other marginal locations were examined also. On July 25 a Boise miner reported a fairly typical situation:

> I will now endeavor to tell you what I have seen and know about Snake river. First, for rattlesnakes, scorpions, musquitos, gnats, sage brush and hot sand, it is the best country I have ever seen; but as for gold and a mining country, I cannot say as much, although there is scarcely a place on the river that a man cannot get a prospect, but not in sufficient quantity to pay; the gold is so fine and light that a miner from other countries, is very easily deceived here. We located a claim, after prospecting about twenty miles of the river, on a piece of graound that we thought would pay $8 per day to the hand but after working it we find it will pay only $3 per day, and this is liable to chop on us any time. There are hundreds of men running both up and down the river that cannot find a place to make grub. They say they don't know what to do or where to go. Some say here that rich mines have been found in the Wind River mountains; others say they have prospected there for the last two years and found nothing.

Some prospectors had better luck. Ralf Bledsoe noted on August 14 that his rocker yield had reached $167 (or $6,400 or so in 1980 prices) in one week and $114 the next. In 1870 prices his return would have set off a major gold rush if more claims of that caliber could be found. Enough miners had come to the Snake that a stage line from Corinne, Utah, began regular service in the summer of 1870 at a modest rate of $15 for a 180-mile trip to the mines.

By 1878 and 1879, when interest revived on the Snake River, new districts below Raft River—at Cold Springs, at Reynolds Creek and Munday's ferry on the Boise-Owyhee road, at Goose Creek (near later Burley)—joined Eagle Rock (renamed Idaho Falls a little more than a decade later), Salmon Falls, and towns along the later Hansen bridge-Shoshone Falls segment as active mining camps. Early in the spring of 1879, a new mining district included the course of the Snake from Raft River to Goose Creek. Another major area at Bonanza Bar, west of American Falls, gained prominence then. By 1882, Boise Valley's New York canal was projected to bring a large volume of water to Snake River placers near the site of Fort Boise, scene of the original Snake River gold discoveries in 1855. When finally constructed, this canal served only for Boise Valley irrigation. Even then, interest in fine gold had led to the design of a large canal system that since 1900 has provided water for most of the Boise project.

By 1880, careful observation of Snake River fine

gold had identified very small particles, so small that five hundred had to be collected to obtain enough gold to equal one cent. Yet at that time, much of the even finer gold—for which three thousand to four thousand particles had to be gathered to recover a penny's worth—could not be recognized at all. Considering that a $5 gold piece was only about the size of a copper penny, those particles had become very fine indeed.

Although perfectly enormous possibilities for the production of Snake River fine gold seemed to exist, if a suitable recovery process could be developed, the whole proposition was entirely too much like salvaging incredible amounts of gold from the world's oceans. Serious efforts were devoted during the next several decades toward finding an economical recovery process for Snake River fine gold.

Unfortunately for the miners, no one ever seemed to be able to solve the problem entirely, although in various stretches of the river, mining operations continued for many years. Altogether, more than 66,000 ounces of fine gold came from the Snake.

BLACK PINE

Silver lode discoveries not far from the Kelton road—Boise's rail connection after 1869—brought a modest array of prospectors to Black Pine at a time when the Snake River fine gold excitement of 1869 accounted for mining interest along other parts of this important stage and freight route. Unlike many of Idaho's gold and silver lodes, Black Pine (an isolated high butte above Raft River) did not suffer as a location remote from rail or wagon transportation. Technological capability to handle complex silver ores still had to be developed at that time, so Black Pine could not profit greatly from the availability of superior transportation. Promotion of an 1870 property there, assisted by additional discoveries the next year, finally brought Black Pine more attention than the area deserved. A Kelton correspondent informed the *Daily Corinne Reporter* on September 26, 1871, of local Black Pine excitement then current:

> All is excitement here about the Black Pine Mines, every speculative and unoccupied man has gone there. We have news that an old location, more than a year ago, has just been bonded for a large amount, and as I write, a citizen of our town has arrived with specimens from a new ledge just discovered some three miles away from the old locations, which he avers is the biggest thing yet, in fact the country is not prospected at all; some time ago, during the Snake River

excitement, Doc. Rice, Mr. Majors of your place, and a few others, made locations but all left them, none worked to develop or explore further, except Rice, who has clung with a tenacity which is now being rewarded; the old locators are hurrying back to save their claims from being jumped. I may take a deck passage on a cayuse, the coming week, and visit the mines, when I can speak more by the card.

Upon returning to Kelton, he described the wonders of Black Pine in a facetious style often employed in mining accounts of the time:

> In my last I told you that I should probably inspect the Black Pine Mines before writing again, and so, one fine morning I started for the hills, distant thirty miles. It would be needless to tell you of the charming alkali plains we traversed, dotted with the picturesque sage, behind every shrub of which peeped a rabbit or skulked a chicken. Suffice it that we arrived before dark at the camp, which is situated at the highest point where water can be obtained. Here we hammocked for the night, putting up at the Hotel de Shively, kept by 'Jim' and his estimable lady. Up at daylight for a climb to the mines which are found a mile above us on top of the ridge, but so steep is the ascent it takes the workmen one hour and a quarter to get up to their labors, which are being prosecuted on the Black Pine and Aerial lodes, owned by Lewis Johnson & Co., said to be an association of English capitalists. After reaching the top of the mountain, which was as near the zenith as I ever expect to be again, we inspected the mines. Work on the Aeriel being done sufficient to answer their contract with the original locators, they have concentrated the forces on the Black Pine, which is a real fissure lode, with a shaft down forty feet; and at fifty feet it is the intention to drift both ways and ship ore steadily, which will be about Christmas. At that time the wagon road will be completed to the mines. That, like all other immensely rich deposits, such as White Pine, Pioche, etc., are only found on the tops of mountains where nothing but silver can grown. Veni, vidi, viei: I cam, saw, I got. You see, Judge, I have not forgotten all the Latin that was flog'd into me. Well, after prospecting around a little while we found the biggest thing out—a monster ledge composed of chlorides, bromides, sulphides and all the other ides, which is richer than pure silver itself; and now, instead of having two good feet (which I have been praying for so long) I have two hundred, and am a millionaire!

After a decade of relative inactivity, Black Pine revived somewhat in 1881. Two or three years of exploration resulted in four small shafts of ten, thirty, fifty, and sixty feet and some prospect holes and cuts. Assays ranged from $28 to $800 in silver carbonate. Alexander Toponce, who had a notable record in Idaho and Montana mineral development, had an interest in Black Pine, but he did not regard his enterprise there as worthy of mention in his autobiography. A fair amount of evidence of mining still remains to be seen at Black Pine; however, this district never attained any great importance in Idaho's mineral history.

YANKEE FORK

Yankee Fork got off to a surprisingly slow start. Joel Richardson and a party of Yankee prospectors examined Yankee Fork while traveling through that part of the country in 1866 or 1867. Aside from bestowing a name on the stream, they left little imprint before retiring to Montana. By 1868, a few men were washing out gold at nearby Robinson Bar. After the rush to Loon Creek in 1869, mining was under way on both sides of Yankee Fork.

Prospectors radiated out in all directions from Loon Creek. D.B. Varney and Sylvester Jordan brought a group of miners over to Yankee Fork in 1870, where most of their claims proved a disappointment. Only one of the new Jordan Creek claims yielded enough (in this case, $10 per man a day) to justify working. The next spring the decline of Loon Creek inspired two more gold hunters to cross over to Yankee Fork. They had a hard time of it. According to Clitus Barbour, "Arnold and Estis [Estes] the discoverers of Yankee Fork camp, toiled in the snow and storm twenty-five days transporting their supplies in there on sleds from Loon Creek, a distance of only twenty-five miles, over a divide thousands of feet high." On the strength of opening discovery claims good for $8 a day, about twenty miners organized a district and went to work. By the end of July, five companies were preparing their claims for mining. Fifty or sixty men, mostly from Loon Creek, were on hand. Some of them "were busy opening their claims, while others were running up and down the river, uncertain what to do, and waiting for the turn of events." Not until the new claims turned out profitable did the doubters go to work. Even then Yankee Fork attracted little outside interest. Only fifteen men spent the winter, and no grounds for a stampede materialized in 1872. Lode discoveries, in fact, did not come on any important scale for three more years.

Searching on a Sunday afternoon in June 1875 for the lodes from which Jordan Creek's extensive, but otherwise unimpressive, placers originated, W.A. Norton came across the vein that every prospector dreams of finding one day. Very few ever had his kind of luck. In a high-grade vein he found a seam of exceptionally rich ore only two or three inches thick. With the help of a partner or two, he was able to pound out $11,500 worth of gold in a hand mortar in thirty days. That was enough to pay some oppressive debts and to start developing the mine. No rush to Yankee Fork attended Norton's discovery of the fabulous Charles Dickens, as it was called. His find went by almost unnoticed. Then, when winter struck early, Yankee Fork was depopulated almost entirely. Packers had no opportunity to supply the high mountain camps. Yankee Fork was reduced to a population of only three, while neighboring Loon Creek declined to four.

⚒ 🐎 ⚒

When prospecting resumed in 1876, other extremely rich lodes followed the Charles Dickens. Most notable of all was the General Custer which James Baxter, E.M. Dodge, and Morgan McKeim discovered on August 17. In a manner somewhat different from the Charles Dickens with its wealth of ore suitable for hand mortaring, the Custer also rated as a prospector's dream. In this discovery, most of the vein happened to lie exposed on the surface. (The way miners describe it, most of the hanging wall simply had slid off the vein.) Thus the miners could avoid much of the expensive development work (that is, driving tunnels and raises or shafts deep into the mountain along a mineralized vein in order to verify presence of enough ore to justify bringing in a mill) ordinarily required before a prospector could sell out his discovery. Erosion already had done most of the development work. Moreover, the relatively low cost of getting out high-grade ore from the Custer enhanced its value greatly. One man could pull down twelve tons of ore a day. E.W. Jones reported in 1877: "The owners merely break the ore loose . . . tumble it down in large masses to the dump, break it up, sort it and sack it." At that point, the ore was ready for packing to a mill in Salt Lake City, where $60,000 was realized from the small open cut. Somehow even this marvelous discovery did not generate an old-fashioned gold rush to Yankee Fork. A complicated and somewhat peculiar claims litigation, involving the original owners as well as two or three sets of subsequent purchasers, held up development of the Custer for a time. The three original locators eventually received $60,000, $105,000, and $121,000 for their claims, depending, apparently, on how expertly they held out.

During the delay, prospecting went on in the locality. D.B. Varney and other old timers from

Custer in 1880

Loon Creek came across the Montana mine along with other promising mines high on Estes Mountain in 1877. Soon they were having high-grade ore packed out for milling. By this time, the Yankee Fork region had enough activity to justify building a town or two. Bonanza City began near the Charles Dickens during the summer of 1877, and another community of Custer followed over near the Custer mine a year or two later. Many of the settlers came from Loon Creek, or from distant Rocky Bar. Utah miners also had good representation in the new city of Bonanza, which showed more promise than growth during its first two seasons.

Lode miners hoped that rich surface deposits would help meet the heavy expenses of developing a major quartz mine, so that they would not have to sell out or bring in large-scale outside capital. Norton's Charles Dickens was one of the exceptional instances in which this actually was possible. From 1876 through 1878, after his highly lucrative initial season, the Charles Dickens yielded about $60,000, mostly through hand mortaring the richest part of the ore. In the first summer, he had $3,400 worth of

rock hauled out on pack mules and then freighted to San Francisco. After his mine was better developed, he had another batch of ore packed out to a mill at Banner. Then in 1878, he sent $30,000 worth of rock (still by pack mule) for milling in Salt Lake City, another $3,300 to Bannock, Montana, and $7,000 worth clear across the Atlantic to Freiburg, Germany, where experts could process it efficiently for testing. In order to avoid such expensive hauls, Norton put in two water power arastras which handled 2½ tons daily. Made of local rock and wood, these inexpensive arastras yielded $33,400 in 1878 and $40,000 more in 1879; in addition, over $25,000 worth of better rock was packed out from the Charles Dickens in 1879. Norton's arastras were losing over one-third of the gold, but others were handling rich ore much more efficiently.

☒   🐴   ☒

A belated—or else premature—gold rush to Yankee Fork finally brought thousands of miners in the spring of 1879. Bonanza City, which could boast only one store and one saloon in the summer of 1878, finally boomed. By the beginning of April, the *Salt Lake Tribune* reported the roads to Yankee Fork to be "lined with stampeders to the Salmon River mines. They are afoot or on horseback, in bull teams and shaky wagons, and to old timers it looks like Pike's Peak and White Pine rush." Writing from Idaho City on July 16, Milton Kelly noted:

> Bonanza City is growing rapidly—as fast as building material can be obtained. There is already a population of two thousand persons in the town and immediate vicinity, all anxious to build and locate permanently. There have been some seven or eight thousand people in those mountains in the present season. Many have left and others are leaving daily, but they are that class who went there to seek employment as miners, and finding the camps new and with little work going on have concluded that they came too early. I have talked with some of these men and they all agree that Yankee Fork is a very rich mining district, but that some time must elapse before developments have progressed far enough to make it a desirable place for those who have to depend upon their labor as miners.

By fall, Bonanza's population had stabilized at several hundred, and times again were dull.

Before large-scale production could be achieved at Yankee Fork, a road had to be built and a major mill installed. More than any other operation, the Custer mine (together with its neighbor on the same vein, the Unknown) justified the erection of a large mill. During the litigation over title, a series of high-priced transactions for portions of the lode absorbed what capital was available. One Salt Lake City commentator suggested, quite reasonably, when one of the purchasers had spent $60,000 for a disputed title to a two-thirds interest, that "sixty thousand dollars is a big price to pay for two-thirds of a law suit; but otherwise the property is cheap at a few hundred thousand, and they may compromise the matter all around."

By the spring of 1879, Joseph Pfeiffer of Rocky Bar had brought in San Francisco engineers and capital, and had arranged purchases enough to enable work and production at the Custer to resume. "People generally thought him crazy" to be investing so heavily in an undeveloped prospect located hundreds of miles from a railroad and on a practically unimproved pack trail "in a wild, sparsely-settled country, surrounded by hordes of hostile Indians." Yet Pfeiffer had recovered his initial $60,000 investment by shipping ore to Salt Lake City, and his California associates, who included George Hearst and the president of Wells Fargo, supplied the balance (over twice that amount) to straighten out title. The next step was to stop hauling ore by pack mule to distant mills in

Calvin C. Clawson (left) at his cabin in Bonanza

Atlanta or Salt Lake; freight costs to Blackfoot, where rail service was available by 1879, ran $100 a ton. Then George Hearst induced Alexander Toponce to build a toll road to Challis, over which stage service to Bonanza commenced on October 3, 1879. Toponce's road made it possible for Pfeiffer's San Francisco capitalists to bring in a thirty-stamp mill for the Custer. In spite of all the excitement, production at Yankee Fork mines amounted to only $420,000 in 1879. Then, "after many unavoidable and tedious delays," the Custer mill was completed at the very end of 1880. Production in 1881 rose immediately to over $1 million. The Yankee Fork mines at last were showing their great potential.

Once the large Custer mill got into operation, production quickly surpassed the total realized from the Charles Dickens—previously regarded more highly—and eclipsed the Montana as well. Managed by competent mining engineers and backed by adequate capital, the Custer ran steadily for over a decade: around $8 million came from this most notable property in that part of Idaho.

⚒ 🐎 ⚒

Both the Charles Dickens and the Montana had a very different experience from their Custer neighbor. Their extreme richness enabled the original prospectors to retain control and mine the better ore by gouging. Most of the ore, while still high-grade by ordinary standards, was not rich enough to take out by primitive methods and pack over the mountains to a distant mill. Here, as in remote mining camps throughout the West after 1866—and especially during the long economic depression that followed the Panic of 1873—large reserves of gold- and silver-bearing rock were discarded by the gougers, who did the best they could with their limited capital and facilities for processing ore. W.A. Norton obtained a half

General Custer mill

General Custer mill in foreground; town of Custer in background (c. 1900)

Joseph Pfeiffer's residence in Custer (1882)

million from the Charles Dickens before he died in Salt Lake City on July 15, 1884. Then his mine was closed pending settlement of the estate. Now it was too late for satisfactory development, and Norton's successors ran into debt trying to make the Charles Dickens pay.

The Montana mine on Estes Mountain met a similar fate. Open to a depth of 500 feet, with only the truly high-grade ore gouged out, the mine had to close in spite of the large blocks of ore remaining. "This considerable development was all done by a horse whim and produced ore to a value of $350,000, every pound of which was shipped to market by pack train at a great expense and still paid a margin of fully forty per cent profit." The original locators who managed the enterprise "spent their profits with a lavish hand while they were coming easy and were unable to properly equip the mine with the necessary machinery when it was too deep for hand work." So they had to shut down and look for a buyer who might go ahead with major investment for development. The original locators

spent until 1904 making a satisfactory deal.

In contrast to these failures, the Custer, which had not been ruined by a little early gouging in getting started, could continue to produce regularly as long as its ore held out. Actually, about eighty percent of the Custer's ore turned out to be in the exposed part of the vein, and the best part of the rock was crushed by 1886.

The financial collapse of the Charles Dickens, marked by a sheriff's sale in Salt Lake City in June 1886, came at a fortunate time for the otherwise unlucky creditors who had to take over the old property. By 1886, expanding railroad transportation and economic development in southern Idaho had made possible a great revival of lode mining in various camps, such as Rocky Bar and Silver City, that had gone through the same kind of experience during the gouging era. British investment in mining properties in southern Idaho expanded greatly in 1886. In July the Salt Lake creditors of the Charles Dickens managed to unload that appropriately named mine onto a new British concern "organized under the leadership of a rear admiral in the Royal Navy."

Capitalized for £250,000 in the beginning, the new British company invested £412,000 the next year so that the Custer mine and mill could be acquired. That way the efficient, thirty-stamp Custer mill could be used to handle the Charles Dickens ore as well. After this merger, the new Dickens-Custer Company, Ltd., had exceptional fortune in the London market: the stock issue was oversubscribed and £1 shares were going for a premium of two shillings in February 1888. Later that year, though, the British investors learned that company officials were engaged in some unsavory manipulation of their Dickens-Custer stock. Worse still, the company operated at a loss totaling £37,000 over the next four years. The problem was not a failure to keep operations going: in 1890 the company had 115 men at work freighting, wood cutting, and mining on Yankee Fork.

After two years of trying to find some other mining region for their investment, the Dickens-Custer Company shut down in October 1892. That ended the initial lode mining boom on Yankee Fork. Stockholders in London, however, were fully as disturbed by the disaster as were the miners of Bonanza. According to Professor W.T. Jackson: "Through ignorance and mismanagement of this enterprise, the British investing public had squandered thousands of pounds of sterling." Once again, large but played-out western mines had been unloaded on an unsuspecting London market.

⚒   🐴   ⚒

Lode mining on Yankee Fork did not revive on any important scale until a new company acquired the British holdings in 1895. Known as the Lucky Boy, this concern utilized the Custer mill in working a major vein parallel to the Custer, although an entirely new wet crushing plant was installed in 1897 to increase recovery. A cyanide plant, added in 1899, allowed the reworking of old tailings for gold and silver. An 1899 tram to the Custer mill also helped greatly. Milling costs had declined from $25 to $2 a ton, so lower grade ores could be worked. Before operations finally halted in 1904, the Lucky Boy had turned out about a million dollars. Modest in comparison to the Custer, this total was double that of any of the other Yankee Fork lodes.

The Lucky Boy had ore that kept its value as miners worked down the vein. When the cost of hoisting up a steep inclined shaft had increased with every foot of greater depth, and finally had become great enough to consume all profits, operations naturally ceased. Hopes that the company would

drive a new low tunnel to resume work sustained the camp at Custer for a time. Moreover, in 1904 when the Lucky Boy quit, the original locators of the Montana lode at last sold out to a group who drove an essential, new 1,800-foot tunnel to strike the Montana vein at greater depth in 1906 and 1907. Nothing much more came of this enterprise however. Perhaps the old owners were just as well off having spent their profits for their own enjoyment instead of developing their mine any further in this profitless way.

Across from the Montana, the Sunbeam Company brought in a new mill in 1904. Attaining significant production in 1907 and 1908, the Sunbeam gained distinction in 1909 as the major producer of Custer County. Enlarging the mill and installing a hydroelectric plant at Sunbeam Dam in 1910 were intended to enable the processing of enormous reserves of low-grade rock similar in many ways to the mineral deposits of Thunder Mountain. The next year the Sunbeam Company had to shut down and forget about their mountain of low-grade ore; thus the expansion and power plant came too late to be of much use. Attempts at dredging Yankee Fork were also undertaken in these years, but major production of this kind was reserved for the future.

Even though Custer and Bonanza gained the noble status of ghost towns over the years after 1912, mining on Yankee Fork still had a future. Limited operations occurred intermittently until the Depression improved the price of gold. Finally, the extensive low-grade placers, which discouraged the original prospectors after 1870, were dredged with success from 1939 to 1942 and from 1946 through 1951. The dredging grossed $1.8 million – greater production than any mine except for the Custer lodes. With this final boost, Yankee Fork was responsible for $12 to $14 million total production in gold and silver. Well over half of this total came from the Custer alone.

Lucky Boy mine

CARIBOO MOUNTAIN

Some placer mining camps were worked out rather quickly, whereas others lasted for many seasons. The richness of the mines did not determine how long they lasted: the length of the normal mining season, usually as long as water was available to operate sluices and other gold recovery equipment, and the difficulty of handling the gold-bearing gravel, along with the amount of gravel to be processed, generally had more impact on the duration of a mining camp. Most placer miners preferred to get rich quickly and to finish working their claims as soon as possible. But mining districts, which could not be exhausted in a season or two, enjoyed greater stability and performance. Cariboo Mountain, with a short annual water season and with deeply buried placers, lasted a long time as a mining center.

Discovered in the summer of 1870, these mines were named for Jesse Fairchilds, known as Cariboo Fairchilds because he had worked earlier in the Cariboo mines in British Columbia. While contrasting greatly in richness with the fabulous buried placers characteristic of Cariboo, B.C., some of the deep placers on Cariboo Mountain were slightly reminiscent of Fairchilds' earlier experience. The Cariboo Mountain deposits showed enough early promise to set off a modest gold rush from Malad and Corinne — the latter a new anti-Mormon freighters' community on the Central Pacific Railway in Utah. Accounts of the excitement in Corinne and of the beginnings of the new mining district, mistakenly identified at first as in Wyoming, came out of Utah early in September:

> Reports reach us of the discovery of very rich gold mines in the district known as Cariboo, in Wyoming. The precious metal is said to be in the form called "free gold," and the richest location about seventy miles east of Soda Springs and near the headwaters of Green River. Of course, the reports give it as richer than anything yet struck in the mountains. A party is going from this neighborhood, and as the distance is not great, we shall probably have authentic intelligence before many days.

A more accurate report was available two days later:

> At last we have some reliable news from this new Eldorado, and from gentlemen not liable to be mistaken. Messrs. Fisher and Lavey reached Corinne yesterday, direct from the mines. Their party of twelve had located and gone to work but a few days before, when a sudden fire destroyed all their provisions but fifty pounds of flour. Three men were at once sent out, making the distance to Ross' Fork, ninety miles, with no provisions but one sagehen. One of the number

returned at once with supplies obtained there, while the two mentioned came on to this city. Their entire company and all they had seen were making from ten to fifteen dollars per day to the man. The area of pay-ground is quite extensive. Quite a number of companies are already on the way there, two of which got lost in the mountains by attempting to find a shorter route, and suffered considerably. The only direct and safe route is to go up the regular Montana road to Ross Fork, from which place a trail leads off a little north of east of ninety miles to the center of the district. A large map, posted up at Ross Fork, shows the exact route. Fisher and Lavey rode horseback from the last point here, over a hundred miles, in a day and night, about the quickest time on record in these parts. They have purchased 10,000 pounds of supplies and several hundred picks and shovels, with which they purpose to make good freighting time back to their locations. The supplies were obtained of Barratt, who worked all of last night to get them shipped, and the scenes around his store this morning remind one powerfully of the old times of 'gold stampedes.' Now that the mines are an established and ascertained fact, whether rich or not, quite a number of Corinnethians are preparing for a start, of whom more anon.

By September 12, the rush to Cariboo provided an excellent vacation opportunity for almost anyone in the vicinity who wanted to see the new mines:

> It is not yet a week since the discovery of the rich gold diggings in Eastern Idaho became known through the towns and settlements along the roads between here and Montana. Thomas Winsett informs us that he was at Malad City when the account reached that place, and in an hour afterward there were parties of from two to ten on their way to the gold fields, and all the way down to Corinne he met people going up to try their fortunes. In addition to the party that left here yesterday, we notice now some more, including many of the business men of the city, who are to start to-morrow. Among these are Harry Creighton, Mr. Burgess, Mr. Short, J.W. Wallace, Geo. Wright, and a number of others. The distance being only a four or five days' journey, and the road a good one, the trip, outside of the nature of the expedition, will be pleasant to the participants. Later accounts all indicate that the district is a basin of great extent and richness. The only practicable route of travel in there is that described in our last issue, namely, the stage road to Ross' Fork, 120 miles from Corinne, and thence 90 miles northeast to the district. We are informed by persons long acquainted with that part of the country that these mines are in Idaho, and not in Wyoming as we inadvertently stated on Saturday. This city is the nearest starting point on the railroad, as well as the most convenient supply depot for the new

diggings, and all present appearances promise that we are destined to have an immense trade this Fall with the miners of Idaho.

Because of the high elevation and lateness of the season, those who had joined the Cariboo gold rush in 1870 could not do very much except prospect when they got there. They could go out panning gold to find the best claims; but with acres of gravel to be worked, panning was too slow and difficult a process to use for gold production. Sluice boxes (in which a strong current of water carried placer gravel over slats that trapped and separated out the gold) worked best. But they could be used only when a lot of water could be brought through ditches to the better claims. Still, some water was available for operating a sluice box, which did not function as well as it should, and the miners gained confidence that they would have a lively camp the next summer.

Reports from Cariboo the next spring continued to show optimism. J.H. Stump, who had a salt mine between Soda Springs and Cariboo, passed on some reliable information on April 24 concerning the new gold camp:

> The first news of the season from the new diggings reached us through Hon. J.H. Stump, of Malad City, who has just returned from Soda Springs. He says that some of the men who remained all winter in the basin, came out a few days ago after supplies, and they had sums of three or four hundred dollars each in good dust, panned during their sojourn. The road in will be passable in a few days, and the placers in McCoy's basin will afford six dollar diggings to several hundred men, with good prospects for still richer product than this. Many miners are now getting ready up the road to go in to New Cariboo, and all are satisfied that a good season is ahead for the summer.

Cariboo Mountain rises to an elevation of 9,803 feet, and most of the mines were found at high elevation. Few other Idaho camps were anywhere near that high. Heavy winter snow prevented much in the way of mining for half the year; and when the deep snow finally melted, water ran off quickly so that, without water, little could be done much of the rest of the time. This situation was typical of mining in the high country.

When spring finally came to Cariboo in May, those who had spent the winter had a chance to give their sluices a better test. Keenan, Allen, and Davis, the pioneer company that started operating in the spring, recovered $60 in a day and a half. Their return ran high enough to encourage construction of a saw mill to turn out lumber for more sluice boxes. This success helped to overcome the discouragement arising from the difficult mining conditions in the new camp:

Since the discovery of gold in California up to the present time every new mining camp has had its rush and stampede of eager fortune seekers, and likewise a reaction. Rushing to the new El Dorados without aim or purpose, except to find gold in lumps on the surface, the excited stampeder is disappointed and soon disgusted if shining nuggets as large as boulders are not as thickly scattered before his view as leaves on the strand. Cariboo has not been an exception to this characteristic custom, and although the mines had hardly been prospected, the most discouraging reports were circulated concerning them, until all confidence had been destroyed in the new region. But a shrewd miner's proverb runs that 'gold is where you find it, and not where lazy men say it is not.' Reliable advices from Cariboo were received here yesterday, which convince us that extensive gold fields exist, and that the yield of dust will be large. A letter received last evening by Mr. Kupfer from a Mr. Meyer, a reliable gentleman mining at Cariboo, states that active mining had but just commenced, and that the most encouraging results had been obtained from several 'clean ups.' Two men working the claim immediately above the writer had made their first 'clean up' the evening before the date of the letter from a ten hours' run and realized $23 and some cents in good dust. Three claims further above the claim of the writer three men had cleaned up $43.60 in one day's sluicing. The yield of some other claims is given, but we have quoted sufficient items to demonstrate that Corinne has valuable gold fields in her immediate neighborhood, which will cause a stir before the dog days are over. Mr. Meyer pronounces the dust of Cariboo as worth more than the average stuff, being fine and pure. The credit of claim owners is said to be good, and a poor man who wants to work his mine can get all his tools and 'grub' on 'tick.' An express will shortly be put on between this city and Cariboo, when information will be received more regularly from the new El Dorado.

Confirmation of problems in mining at Cariboo, as well as of the potential wealth of the district, came with a report that the miners wanted to import Chinese men to work the placers:

> A gentleman now here from the gold diggings of Cariboo, reports to us that claim owners are steadily making $20 a day to the man. Ground on McCoy and Iowa creeks, is growing richer, but there is great scarcity of hands. Three hundred men could find work by the day, now, at $4, but those going in generally stake out claims and work them in preference to taking wages, and hence the drawback in securing labor by those who own the best paying placers. Water is abundant in all the streams, the roads into the mines good, and many miners from Idaho and Montana are gathering there. An effort is being

made to obtain Chinamen to work the diggings, as white men are not to be found in sufficient numbers to supply the demand for laborers. It is somewhat astonishing that gold placers of such extraordinary richness, only two days' journey from the railroad, should thus go a begging for men to come and gather up their wealth, but this is the actual fact. If any one would see the evidence of Cariboo's richness, let him drop in at any of our banks during the day and see the sacks of dust which miners are exchanging for coin and currency.

As was the case in most mining camps that had gotten over the hard feelings engendered by the Civil War, Cariboo miners had grand, old-fashioned Fourth of July celebrations even though they had to hold their parade in deep snow.

The true spirit of patriotism was unmistakenly and forcibly evinced by the miners and citizens of Carriboo and Soda Springs, on the Fourth [1873]. In Carriboo the usual exercises were gone through with, such as cannonading, bell-ringing, fire cracker shooting, cocktailing, etc., a grand oration, a procession on snow shoes, a huge feast of bear meat, trout, mountain sheep, grouse, and other luxuries of that district, a war dance in the afternoon by a few of the noble red men residing in those parts, and a grand ball in the evening, which wound up with a roaring serenading party that browsed round on the rim of the basin to the tune of 'Hail Columbia.' Thus ended the Fourth at Carriboo, but quite a number of the miners and merchants living there went over to Soda Springs to celebrate, and there, in the shade of the veteran pine forest which covers those medicinal founts with its umbrageous foliage, they lavished multiplied encomiums on the heads of the sires of '76, while a stream of pure sparkling soda water gently disappeared between their patriotic lips. A glorious soiree also wound up the day's programme at Soda, and a day never to be forgotten by the participants at these two places wore pleasantly away, and slipped, reluctantly, away into the never-ending eternity.

Aside from requesting Chinese workers to help, miners from Cariboo resorted to labor saving equipment that was standard in many western camps. Hydraulic giants were installed to obtain placer gravel to feed the sluices. Powerful streams of water (shot out of nozzles fed by metal pipe leading from ditches at higher elevation) cut away surface gravel and swept the gold-bearing placer gravel into sluice boxes. Within a year or two a number of hydraulic giants were at work in the region, and by the fourth season, eight of them had gone into production.

⚒ 🐴 ⚒

As the years went on, gold recovery at Cariboo proved erratic. A few spots yielded well, but most of the ground turned out to be marginal. One or two claims gave satisfactory results—an ounce a day (about $20) for each miner at work. Most of the others provided from $2 to $5, with the leaner ones of interest mainly to the Chinese. Unlike miners in most Idaho districts, those at Cariboo made no effort to exclude the Chinese. They never seemed to get enough white miners to work the ground available, so driving out the Asiatics seemed pointless. Chinese companies owned claims and operated giants along with everyone else—apparently without discrimination. Unlike other camps that forcibly kept out Chinese competition during the more productive early years, Cariboo had whites and Chinese at work on adjacent claims most of the time and did not become strictly an Oriental camp after a few seasons of early activity.

In the early days of placer operations, Cariboo Mountain had two mining districts, one on the east side at Iowa Bar and the other on the west with Keenan City as its center. Keenan City, with a dozen or so log cabins on McCoy Creek, had become the major (and only recognizable) mining center. Iowa City on the other side of the mountain was pretty hard to find, even for the people who happened to go through it. In the Iowa district (named for an Iowa discoverer), William Clemens —a cousin of Mark Twain who mined in various parts of Idaho for more than thirty years—spent many seasons placering and promoting the country. He had three hydraulic giants in operation, the most in the district. Cariboo Fairchilds spent fourteen years in McCoy district not far from Keenan City and had quite a time: In 1872 "he broke his leg while 'skylarking' with a friend one day," and in 1884 he had a fatal misadventure with one of Cariboo Mountain's numerous bears.

With the discovery of lode claims in 1874, Cariboo Mountain offered an additional attraction to early miners. Over the next decade, a number of these new lodes were developed to enough depth to prove that thousands of tons of ore were available, if anyone would operate a hard-rock mine in such a remote and difficult location. Simple arastras— rock crushers made of local materials, with drage stones used to grind up ore in a circular rocklined surface—provided modest production. But not enough ore in thousands of tons was available in one place to justify a major stamp-milling enterprise in the early years, although such a possibility attracted attention to the district season after season.

Even though the slowly worked placers proved spotty, with an occasional rich streak at bedrock, by 1886 production may have amounted to $1 million. Reports of $200,000 in 1879 alone suggest an eventual total that large—or perhaps twice that much if enough of the other seasons provided as much as $100,000. However, considering the relatively small number of miners at work most of the time, and the shortage of water high on Cariboo Mountain, even a $1 million total is difficult to substantiate. All kinds of exaggerated reports of mineral wealth came out of most western mining camps, but with enough short seasons and a fair number of hydraulic giants at work, Cariboo had provided a substantial return for a modest number of miners.

Remote from other mining districts and distant from sources of supply, Cariboo Mountain provided a definite economic stimulus to early development of the upper Snake River country. At least one enterprising miner found that he could grow premium Idaho potatoes next to a snowbank high on Cariboo Mountain at a time when few farmers were at work in the valley below. But generally the miners at Cariboo had to depend upon distant sources of supply, and their needs offered an inducement to settlers to develop the surrounding country at a time when not too many other economic attractions were available to encourage settlement in that part of Idaho.

Rail service to the broad valley of the Snake River to the west eventually helped in the development of Cariboo Mountain lode properties; and about the time Idaho became a state in 1890, a long awaited stamp mill served the Robinson lode. A pretty good test of the only developed property was made. From a 246-foot crosscut tunnel, ore was removed from a 275-foot stope on a 25-foot vein opened to a depth of 264 feet. The mill burned in a fire not long after that initial orebody was processed, and the company did not bother to replace it. Edmund B. Kirby reported in Denver on August 6, 1894:

> A great number of quartzite strata near the top of the mountain have been prospected by surface pits, and are found to be gold bearing. The surface soil and gravel down the entire slope of the mountain on this side is said to pan well in gold. These strata range from three to sixty feet in thickness.

Geologically, these gold deposits on Cariboo Mountain had the same corrosive alkaline origin and structure that characterized quartzite gold deposits in Ouray and Battle Mountain, Colorado.

After years of sporadic effort, several companies gradually made headway in trying to develop lode properties on Cariboo Mountain. An Idaho Falls corporation, established on October 8, 1903, invested $60,000 in an 800-foot tunnel and a 100-foot shaft on the Monte Cristo. Copper discoveries in 1904 encouraged the Monte Cristo

investors for a time, but the deposits could not be mined successfully. Eventually a Salt Lake City enterprise was incorporated on April 19, 1917, to spend another $13,000 on the Robinson lode, which finally attained 1,200 feet of tunnels — compared with only 246 feet during the era of production. In a still more ambitious project undertaken by a Boise company incorporated on March 27, 1920, four men drove 1,600 feet of tunnel for the Searchlight. But that concern soon shut down too. After many years of idleness, activity resumed at the Robinson in 1938, when leasors put five men to work. Eventually another 100 feet of tunnel was driven in 1952, and more prospecting followed on the Robinson in 1955. Work at the Evergreen also helped maintain interest in the district during those years. Although a ball mill was utilized after 1940, large-scale lode mining simply could not be managed on Cariboo Mountain.

Twentieth-century placer mining did a little better. With ten men at work in 1907, the American Placer Company handled about 50,000 yards of gravel that season. Then an experiment in dredging came with the efforts of the Wolverine Placer Company, incorporated on May 10, 1917. With eight men and capital outlay of $70,000 to install a 150-horsepower hydroelectric plant, eighteen miles of transmission lines, and a McCoy Creek dredge, this operation showed promise before work was suspended in 1922. At that time, an ambitious placer operation on Barnes Creek was started with Pittsburgh capital. A four-mile ditch fed water into a 2,500-foot pipe that finally supplied two four-inch hydraulic giants and a 300-foot sluice line. A substantial permanent camp accommodated twenty miners. A thirty-horsepower hydroelectric generator with a mile transmission line provided power for a modern amalgamation plant. A warehouse, blacksmith shop, assay office, boarding house, and bunk house, along with a company office building, formed part of this ambitious placer operation. When the Wolverine dredge resumed production in 1924-1925, Cariboo looked still better.

As with most old gold camps, Cariboo profited by the revival of interest in gold mining during the Depression. A Minneapolis company, incorporated on May 3, 1936, employed four men to move 16,000 yards of gravel in 1938. Small operators, able to engage in subsistence mining on old placers, recovered modest amounts of gold, enough so that they could stay off government relief, throughout most of the Depression. But total production from this Minneapolis company enterprise amounted only to $4,000 to $5,000 a year in 1938 and 1939, so the addition to the district's output still was limited. From the time that George Hearst had sent experts to examine Cariboo Mountain in the fall of 1879, local promoters anticipated that "the next year of our history will show a record of population and mineral wealth unparalleled in the records of our Territory." But somehow Cariboo's golden age never materialized.

BAY HORSE AND CLAYTON

In 1864 a lone prospector travelling through the Salmon River Mountains with two bay horses found excellent mineral locations along a stream soon to be named for his steeds. Descending to the Salmon River, he met a party which had just come down from Stanley Basin. Encouraged by his discovery they examined Bay Horse Creek without success. Yet a search for Bay Horse lodes went on. Finally W.A. Norton succeeded in finding a mine in 1872, and he and S.A. Boone located a lode on September 1, 1873. Later, A.P. Challis and a group of prospectors turned up a series of promising veins at Bay Horse in the spring of 1877 and found enough silver values to cause a considerable rush there in 1878. After that, Challis took over the Norton and Boone property. Lead-silver mining in Nevada and Colorado had reached a stage of development sufficient to generate interest in similar lodes in Idaho. Within a year or two, even more spectacular lead-silver discoveries brought a major rush to Wood River. Idaho mining had entered a new era of development.

Of a number of interesting claims, Tom Cooper and Charley Blackburn's Ramshorn offered superlative prospects. Unable to develop such a remote property, Cooper and Blackburn sold for a modest price to a group who managed to have thirty tons of ore sent out to Salt Lake City in 1878. Fortunately the new owners recovered $800 a ton from their initial shipment. In 1879, they did still better, adding about $70,000 (or maybe only $32,000) to their total. Even with seventy-five tons running at $900, they had to spend $47 a ton on transportation. Using pack mules to get the ore down Bay Horse Creek, they employed wagons for freighting to Blackfoot after the railroad reached that new community in 1879.

Other Bay Horse lodes showed similar high assays. Robert Beardsley and J.B. Hood funded additional development work with a twenty-two ton shipment from their decidedly superior property in 1879. Eventually their lode almost matched the Ramshorn. These two properties accounted for the greater part of early lead-silver production at Bay Horse. Several other smaller shipments attested to additional lode values around

Bay Horse in 1879. A 7,228-pound lot from a lode above Clayton returned a profit of $550 that summer. Additional properties showed promise at Poverty Flat. Smelters at Bay Horse and Clayton (named for J.E. Clayton of Atlanta, who selected the mill site) clearly were justified by these test lots. New owners of the Ramshorn (including E.W. Jones of Idaho City and A.J. McNab of Salmon) managed to arrange for additional investment to provide local reduction facilities, and Omaha capital was induced to support Clayton's smelter in 1880.

Both Bay Horse and Clayton gained stability and population with the construction of thirty-ton smelters in each camp in 1880. Initial operating expenses ran high for these pioneer Idaho smelters. Coke had to be imported all the way from Pennsylvania to operate them. By 1882, however, charcoal, a substitute for coke, was being prepared locally. Forty-eight men were employed in providing 180,000 bushels of charcoal to maintain smelting operations. When smelting resumed June 14, 1882, after a couple of short previous seasons, more than enough ore was available to maintain continuous production. Over $300,000 in silver alone came in 1882. Lead production increased the output to 2½ tons of bullion daily from twenty-three tons of ore. Bay Horse had gained a population of about three hundred, and boasted having a complex of substantial, permanent buildings—mostly saloons. A meat market, a general store, and several boarding houses added variety to the growing community. John T. Gilmer and O.J. Salisbury (prominent stage line operators) acquired the Ramshorn in 1882, enlarged their Bay Horse smelter, and erected a thirty-stamp mill late in the year for their Bay Horse property. Then on September 1, 1883, they opened a major Ramshorn tramway to transport ore down to wagons which served the smelter. With plenty of ore on hand,

Bay Horse charcoal kilns

Bay Horse

their operation assured Bay Horse a bright future.

Even though they had a successful smelter at Bay Horse, Gilmer and Salisbury soon faced a problem in mineral recovery. Much of their lead came from a lode which lasted only four years. So in 1884 they had to start hauling Elkhorn lead-silver ore from Ketchum to Bay Horse, an interesting reversal of their usual direction for shipping, in order to provide sufficient lead to gain efficient operation. Although a variety of ores available nearby contributed to the diversity necessary for effective smelting, Bay Horse (like many remote lead-silver mining districts) lacked enough lead to maintain a proper composition necessary for mineral recovery. This problem, characteristic throughout the west, favored large smelters adjacent to transportation centers, which could receive ores from many different mines and mix them to obtain essential chemical reactions. Others might need a component different from additional lead, but most suffered from one shortage or another.

While Bay Horse flourished, Clayton's thirty-ton smelter supported a more modest camp. Only a half dozen families lived there in 1882, but two general stores and a saloon did open for business. A modern thirty-ton smelter occupied a sixty-foot by one hundred-foot building which boasted an iron roof. Without a mine the size of the Ramshorn, Clayton's smelter ran for a shorter summer season. Each summer, Clayton's production resumed with enough variety of mines to maintain a proper balance of ores.

After Gilmer and Salisbury solved their problems in smelting lower grade ore in 1887, Bay Horse continued to operate at a capacity restricted primarily by inadequate transportation. In 1888, Bay Horse's last full year, more than 150 miners supplied ore for milling, concentrating, and smelting. Ten men operated a mill, another ten ran a concentrator, and twenty-five handled a smelter that could have been doubled in size, if their increased product could have been hauled out. Four tons of bullion, worth $750, came from about thirty tons of Ramshorn ore processed each day. Silver values ranged from 80 to 1,800 ounces, while lead made up most of the four tons turned out daily. An adjacent property, the Skylark, provided about sixty tons of ore for smelting daily, with a total yield almost equal to Ramshorn production.

A series of misfortunes plagued Bay Horse in 1889. A fire on May 14 set back operations, destroying Sing Lee's wash house ($1,000), a stage company barn ($1,400), Charles Small's teamster's dining hall ($800), and two $1,500 dwellings. A water shortage slowed production in August. Then a change in federal tariff policy, allowing the importation of Mexican lead, precipitated an abrupt shut down at Bay Horse in November. Declining silver prices aggravated the situation. In places such as Bay Horse, silver values in lead-silver lodes enabled miners to produce lead at substantially less cost than ordinary lead mines could manage. So lead prices declined, giving places like Bay Horse and Wood River a competitive advantage. But when Mexican imports from similar lead-silver districts threatened to reduce lead prices still further at a time when silver profits were declining, Bay Horse had to close until transportation costs could be reduced or until prices might increase. O.J. Salisbury started up his Ramshorn smelter for a month in 1893 and tried again in 1894. But after that he got his smelting done in Clayton. His Ramshorn mine had accounted for $2.5 million—around a quarter of Bay Horse's $10.25 million early production. Bay Horse's total production included about $6.9 million in silver, $2.7 million in lead, and $650,000 in copper. Salisbury soon had about that much more ore developed. All he needed was a more favorable cost ratio to justify the resumption of large-scale Bay Horse production.

Clayton

Clayton fared a little better after 1889. About sixty men worked there early in 1889 before prices got too discouraging. Mines around Clayton supplied a proper variety of ore so that only coke had to be imported for smelting. Late in 1888, Clayton's smelter was enlarged to a capacity of sixty tons. Within two years, smelting at Clayton increased enough to employ a larger staff and to encourage R.A. Pierce to move his Challis newspaper, the *Silver Messenger*, to Clayton for a time.

From 1892-1902, Clayton's smelter managed a productive run for a little more than a hundred days each summer. A small part of the ore came from O.J. Salisbury's Ramshorn development work, which he financed by sending out just enough ore to maintain his mine and to pay for exploration. By 1900 he had six miles of underground development completed, primarily 35,000 feet of tunnel. The location of the Ramshorn in a deep canyon allowed him to gain a vertical depth of 3,000 feet in development with only a 3,000-foot-long crosscut tunnel, instead of a 3,000-foot shaft. Salisbury's enormous low-grade Ramshorn ore reserves did not contribute much to Clayton's smelter, but enough mines closer on Poverty Flat (which yielded about $1.5 million worth) and at other handier locations kept Clayton's smelter in operation. About thirty miners worked each winter providing ore for another fifteen men to smelt each summer. Twenty-five more kept busy in the charcoal camps supplying fuel. Another ten men and fifty horses hauled ore and charcoal to Clayton to keep this operation going. Aside from four thousand tons of coke that had to be imported from Ketchum each season during earlier years, this process did not have to rely upon resources outside the district. Even from 1900 to 1902, Clayton's bullion production averaged over a million pounds each year (1,307,399 in 1900; 954,775 in 1901; and over a million in 1902 in lead) with substantial silver values (109,248 ounces in 1900) in addition. While this represented a decline from 1,426,551 pounds of bullion in 1898, Clayton's summer operations provided a significant contribution to Idaho's mineral production for more than a decade after Bay Horse shut down.

The suspension of activity in Clayton after 1902 resulted from the lack of available ore. Although some Bay Horse leasors shipped out 480,000 pounds of lead and 20,000 ounces of silver in 1903, any serious revival of activity was delayed until 1912 when five Bay Horse mines accounted for $67,000 in lead, silver, and copper. A new company assumed control of most properties in 1912, but the minor efforts of leasors comprised what little activity occurred from 1902 to 1918. Then limited shipments of Ramshorn ore resumed in 1918, and a modern flotation plant was completed on December 1, 1919. A monumental retimbering and reopening job preceded the low-grade production of Ramshorn ore that Salisbury had developed more than two decades before. Several years of capacity production allowed Bay Horse to realize a long anticipated $2.5 million return until low water retarded operations in 1924. Finally leasors took over again when Ford Motor Company efforts to revive the Red Bird in 1924 came to a halt in October 1925.

☨   🐎   ☨

After another decade of inactivity, Clayton's smelter started another long period of production in 1935. Rising silver prices stimulated mining to a level of $2.34 million a year in 1968. By that time, Clayton's revival had more than matched earlier production. As southern Idaho's only important silver mine during most of that time, Clayton gradually built up a total production for the Bay Horse region to substantially more than $40 million in lead (over $12 million additional), zinc (almost $7 million more), and silver ($2 million more) in two decades. Shortly before 1960, a consolidation of Bay Horse mining claims was arranged in order to facilitate a Bunker Hill and Sullivan lease option, but after an extensive investigation, that proposal was declined.

After more than four decades of steady production, Clayton benefitted from major new discoveries in 1978 in existing workings (at their 1,100-foot level) that provided an increase in ore reserves from 100,000 to 410,000 tons in 1980. Aside from good long-range future prospects at Clayton, the resumption of drilling in important Bay Horse properties in 1979 offered hope for inactive mines there.

A far more extensive operation commenced on Thompson Creek northwest of Clayton, where development of a major molybdenum prospect began in 1980. Designed to increase world production by 19 percent, ore sufficient to last twenty years at that rate was blocked out. With 200 million tons of reserve for an open pit that would become Idaho's largest, a daily production of 20,000 to 25,000 tons was anticipated. Site preparation and construction of accomodations for 550 employees with an $8 million payroll was completed for initial production in 1982. A $350 million investment to initiate this enterprise had a substantial social and economic impact upon Challis as well as other local communities.

HEATH

Coming north from Nevada in 1874, James Ruth and T.J. Heath discovered four silver claims above Goodale's cutoff along Brownlee Creek that October. They located a millsite at Ruthburg, two miles below their major property, the Belmont, and had ore sufficiently rich to create substantial interest the next summer. William A. West came in from Placerville and added more excitement when he located the Fairview lode on May 8, 1875. A mining district was organized at Heath on June 28.

The identification of copper ore added to Heath's mineral possibilities. Lodes eighty feet in width looked good, although no testing could be done on site to establish the value. A batch of high-grade samples was packed to Rye Valley for assays, and in 1876 a two-ton test lot was assembled for shipment to San Francisco. Eventually the Belmont came to George Hearst's attention. In June 1881, he came up to Boise to purchase the property. A ten-stamp mill, the one to be depicted later on Idaho's state seal, was installed to process ore developed by shallow tunnels. Perhaps 50,000 to 60,000 ounces of silver resulted from this operation. Then lead values increased just enough to upset a chlorination system employed to process Belmont ores, so production had to be suspended. Finally an effort to develop a much larger orebody failed during the Panic of 1907. A six hundred-foot tunnel project, designed to reach a major vein at a depth of four hundred feet, was halted for lack of funds only fifty feet short of the goal. A similar failure attended the development of another property higher on Cuddy Mountain. A camp for seventy-five miners was constructed in 1906, three miles of access road was built, several hundred feet of tunnel was driven, and two hundred tons of fifteen to twenty-five percent copper ore was recovered before the operation was shut down. Then, after an Oregon Short Line extension past Brownlee ferry brought the railroad within six miles of Heath, a much larger copper property attracted a syndicate of Utah and Nevada investors in 1909. Occasional work there, as well as on the Belmont which the Bunker Hill and Sullivan mines explored under lease in 1920, followed over many years. But these properties have not matched early expectations.

After a long period of inactivity, an important silver discovery near Cuddy Mountain in 1977 brought renewed interest to this area. Eight major companies became active there following reports of a lead-silver-zinc orebody with four to eight ounces of silver per ton, supplemented by a half percent zinc. In 1978, additional exploration indicated another billion ton mineral deposit with potential commercial possibilities. So after more than a century, T. J. Heath's mining region was gaining more attention than ever before.

SEVEN DEVILS

So little was known of lead and copper prospecting in the West before 1869 that during Idaho's gold rush era other mineral discoveries, aside from some silver properties, could not be exploited commercially. Rich copper lodes in the Seven Devils country were discovered about the same time that gold was found in Boise Basin, but all efforts to introduce copper recovery technology failed for many years. Gold and silver mining continued long enough so that lead-silver and copper prospects could be developed as part of a western mineral empire in which problems of lode mining and metal recovery gradually were solved while processing traditional metals.

Levi Allen, who originally found and promoted several Seven Devils copper mines, set out with a boatload of fourteen prospectors from Fort Walla Walla (an abandoned Hudson's Bay Company post located where the Snake River enters the Columbia) on March 14, 1862. Aside from searching for new mines, he had an assignment from the Oregon Steam Navigation Company to report upon the practicality of extending steamboat service from Lewiston up the Snake. (His conclusion, confirmed by a follow-up expedition in the fall, that steamboating up the Snake would encounter no serious obstacles — compared, at least, with other northwestern steamboat routes — may have been distorted by the exceptionally high water levels in the spring. Water discharge down the Snake in 1862 greatly exceeded that of any subsequent flood. His party faced the same deluge which had impeded the discovery of gold in Boise Basin in the summer.) After taking his boat to the Salmon River, he further concluded that steamboats could not provide service to Slate Creek to accommodate the great 1862 gold rush to Florence. So he set out in a small four-man party to ascend the Snake to Fort Boise. Reaching the Weiser River, he examined the Payette country and checked out a route to the new Boise Basin mines. Before he returned to Lewiston, he had also discovered spectacular Seven Devils' copper outcrops near Kinney Point, high above the Snake River opposite Hells Canyon Creek. He spent the next quarter century trying to promote his copper claims.

Impeded by the remoteness of his mines and the lack of interest in copper, Allen made no headway until after T.J. Heath and James Ruth discovered the Belmont mine above Goodale's cutoff to Brownlee ferry in October 1874. Commencing as a silver camp, Heath's district attracted considerable interest. By the next summer, large copper lodes were identified. In 1876, substantial promotional

activity gave Snake River copper possibilities more promise, and Allen managed to dispose of a fourth interest in his Peacock mine to Granville Stuart and S.K. Hauser (later a governor of Montana) for $1,500. Hauser sent Isaac I. Lewis to examine Allen's property in 1877, and in 1886, Lewis (a leading businessman in Ketchum who had also acquired an interest in Allen's Seven Devils discoveries) disposed of his Peacock holdings to Albert Kleinschmidt, a Helena merchant. Kleinschmidt also acquired some additional Seven Devils copper properties on Indian Creek, and after some good high-grade ore was packed out from that more accessible lode for test smelting at Anaconda in 1887, preparations were made for serious development of both areas.

Kleinschmidt's Peacock property had a "great surface display of mineral" over an area 550 feet long by 80 feet wide. Some even richer, though smaller, prospects occupied another part of a three-mile-wide mineralized zone that extended about five miles north and south on Monument Mountain. Kleinschmidt went immediately to work to solve the transportation problem: without access to a railway, even a relatively rich copper mine could not be developed. Surveys for a Weiser-Salmon Meadows railroad, which might have run a branch to the Seven Devils, and for a line down the Snake River from Huntington to Lewiston had been made after the Oregon Short Line and the Oregon Railway and Navigation Company commenced transcontinental rail service through Weiser and Huntington at the beginning of 1884. But a boatload of surveyors had been lost in the Snake River, and even the Weiser-Salmon River project faced difficult enough terrain, so that construction was delayed for more than another decade. Meanwhile, Albert and Reinhold Kleinschmidt engineered a $20,000 wagon road — the well-known Kleinschmidt grade — constructed in 1890-91 from the mining district down to the bottom of Snake River Canyon. There they hoped that their mines could be served by a steamboat. However, the steamboat *Norma* built for that purpose in 1891, failed to meet their need. Then the Panic of 1893 set copper mining back for several years, and litigation over control of Kleinschmidt's mining company went on for a decade or two.

Adding to these setbacks, a smelter, built at Cuprum in 1897, failed to work in 1898. A new company then purchased the Peacock and other Monument Mountain claims for $1 million. The beginnings of railway construction promised to solve the transportation problem; a line from Huntington down the Snake to Homestead and the Pacific and Idaho Northern line from Weiser

toward Council began to bring railway service
fairly close on either side of the Seven Devils. While
the rails were advancing toward the mines, thirty-
five carloads of high-grade ore consigned to a New
York smelter were hauled by wagon for sixty miles
to the railroad.

J.H. Czizek, state mine inspector, happily
reported in 1899:

> For the first time in the history of the Seven
> Devils a systematic and thorough effort under
> good management was made to show the vast
> wealth in the copper in this district. The Seven
> Devils district undoubtedly has some of the best
> copper prospects in the United States, if not in
> the world. It is destined at no distant day to rival
> Butte, Montana, in the wealth and productive-
> ness of its copper mines.

Limited production continued in 1900. The next
year the building to house a smelter was erected ten
miles from Weiser to accommodate the district. But
no smelter was installed, and the building served
only as a cattle barn. After the Pacific and Idaho
Northern began to grade a right-of-way toward
Cuprum, the line finally went past Council, but
only as far as New Meadows. Litigation among
various leading Montana mine owners, who had
Seven Devils interests, shut down the district
entirely in 1902. After insolvency of a major
property was liquidated by a sheriff's sale in 1902, a
little ore was shipped out in 1903. Established
copper companies in Utah, with mines at Bingham
Canyon, considered expanding into the Seven
Devils, but the large, economical open-pit type of
operation they had pioneered in Bingham Canyon
could not be utilized in the remote Seven Devils.

Neither of the railways got close enough to
provide economical transportation of ore to distant
smelters, and a smelter at Homestead failed in 1903.
So a smelter that no longer was needed at Mineral
was moved to Landore, right in the central part of
the copper district. An attempt to run the new
smelter on wood failed in 1904, and difficult
transportation ruined efforts to supply the Landore
smelter with imported coke. Some of the best ore
was hauled down the Kleinschmidt grade and
smelted at Sumpter, Oregon — this mainly in 1906.
By this time, R.N. Bell, state mine inspector,
reported that the Seven Devils "district has been
badly handicapped since its discovery by title litigation and some of the rankest kind of mining mismanagement. A large amount of capital has been
expanded on several different properties in the
camp, but without definite results in the way of
intelligent development." But Landore was a thriv-
ing community of a thousand people with daily
stage service to Council, and the outlook seemed

A.O. Huntley built this mansion near Cuprum from his
Thunder Mountain profits

good. About $750,000 was produced from
Landore's smelter and more distant refineries by
1908.

Undeterred by their inability to operate around
Landore without better transportation, the Kleinschmidt interests took an option on the Red Ledge
(ten miles north of Landore) in 1906. The Red
Ledge was described in 1906 as being a mile long
and two thousand feet wide "and as red as a freshly-
painted barn" because of oxidized iron that colored
the copper ore. Hope that the Union Pacific might
build on the Snake River from Homestead (seventeen miles above the Red Ledge) to Lewiston
encouraged them; such a railroad would have
passed right by their mine above Eagle Bar. But rail
construction got only to Homestead in 1909. Finally
in 1925, Idaho and Oregon provided the Ballard
Landing Bridge across the Snake River at the base
of the Kleinschmidt grade below Homestead, and
the Red Ledge owners extended a narrow road on
down the canyon to Eagle Bar in 1926. A fraudulent
promotion of the Red Ledge, along with the properties at Landore, set back the district after 1926.
Operations in all the major Seven Devils mines were

Landore, early twentieth century

suspended on November 4, 1927, and the companies remained in receivership until after the perpetrator was convicted in Federal court on December 14, 1928. These calamities, however, only intensified the problem of developing the district—a problem that sprang primarily from the lack of transportation, which has still thwarted Seven Devils copper mining for over a century following the original discoveries in 1862.

Intermittent work continued in the Seven Devils, with most of the extremely high-grade ore hauled out to smelters. Most of the potential ore has not been touched, even in the mines that have been worked, and the Red Ledge as yet has really not produced anything at all. By 1960, more than $400,000 in copper and much less in gold and silver had been recovered at an overall cost of well over four times the total production. Thus the Seven Devils mines were aptly named: in spite of energetic efforts by a number of important and well-established copper mining companies, the Seven Devils could not be persuaded to release its wealth of buried treasure, and only the surface of the district has yet been scratched.

Meanwhile, production has continued. Leasors on the Peacock shipped out $150,000 worth of ore from 1960 to 1963, and another 27,000 tons of eighteen-percent copper ore with $3.25 per ton of silver and gold in 1966-67. In the summer of 1968, ore trucks were hauling rock from the Peacock down the upper Kleinschmidt grade to Cuprum, then on to Council, Grangeville, and Anaconda,

over a different route but to the same place where the original Peacock ore was smelted in 1887.

Finally, a large open-pit Silver King property above Cuprum provided ore for a new Copper Cliff mill there through 1980. Ore from a nearby Oregon mine was then hauled up the Kleinschmidt grade for processing at Copper Cliff. Then in 1980, this Copper Cliff-Iron Dyke operation was enlarged by a $1.5 million Red Ledge purchase. After more than a century of effort, Seven Devils mining became a large-scale industry.

GIBBONSVILLE

Commencing as a northern extension of a long series of Lemhi placer discoveries, some Anderson Creek finds in the summer of 1877 proved to be too low-grade to gain much attention. But as soon as they were traced to a promising lode in September, an arastra went into production that fall at Gibbonsville. Two more followed in 1878. Enough miners came to justify opening a post office on April 8.

Unlike most other lode mining camps, Gibbonsville was developed by the original discoverers. They managed to cover the cost of getting into operation from the proceeds of their production. Simply working down from their outcrop, they took in enough profit to bring in a ten-stamp mill in 1879. Development funded that way was necessarily slow, but by 1880 Gibbonsville had gained a population of around 175 and attracted enough favorable notice in Butte that substantial British capital was brought in the next year.

With British investment, Gibbonsville took on a more conventional development. By 1882, fifty miners were employed, although the company got into disrepute by neglecting to pay them. Soon the camp was shut down because of litigation. Aside from this problem, Gibbonsville had important natural advantages as a mining camp. Located near the continental divide and close to the Montana boundary, Gibbonsville had a good wagon road through nearby Big Hole to the Utah and Northern Railway. Later, a Northern Pacific branch line from Missoula up the Bitterroot came within thirty-five miles of Gibbonsville. Eventually, the Gilmore and Pittsburgh railroad reached Salmon in 1910. By that time Gibbonsville had gone through more than one phase of development, all of which had been more practical because of available wagon and rail transportation.

Whereas lode mining made Gibbonsville into Lemhi County's major gold producer after Leesburg went into eclipse, Dalonega Creek and other

Gibbonsville mine cars, boiler room, and compressor plant (1912)

Gibbonsville after 1920

nearby streams finally provided significant placer production. Hughes Creek placers followed in 1895, with Minnesota capital to fund the operations. By that time Gibbonsville had become a town of seventy-five to a hundred buildings, with a roller mill and three stamp mills, two saw mills, two stores, and six to eight saloons. A thirty-stamp mill brought in there in 1895 ran a little over two years before the company went insolvent in 1898.

Additional capital investment in the summer of 1898 raised the total number of stamp mills to five, and Gibbonsville's newspaper, the *Miner*, was available to publicize the area. Then, upon reaching greater depth, Gibbonsville's lode turned to sulfide ores which no longer were free-milling. Thereupon, the camp declined abruptly. In September 1899, George M. Watson noted:

> Gibbonsville is a thing of the past. There is not enough ore in the camp to run five stamps, everybody is leaving for new fields and the camp is nothing more than deserted houses and shacks.

This misfortune left Gibbonsville inactive for six years.

A later operation renewed mining at Gibbonsville in 1906, and in 1908 a twenty-stamp mill was constructed to handle sulfide ores. Although almost all production came from extensive low-grade ores, Gibbonsville accounted for about $1.5 million in gold up through 1898. Total production of about $2 million finally made Gibbonsville into one of Idaho's more substantial gold camps. Drilling for uranium at Gibbonsville was undertaken in 1979 in an effort to diversify metal resources there.

VIENNA AND SAWTOOTH CITY

Immediately prior to the Bannock War of 1878, Levi Smiley left Challis late in May to prospect the upper Salmon River above Stanley Basin. Just before his party crossed a high divide onto the South Fork of the Boise, they noticed a rich quartz outcrop. When they were about to record their claims, the news that the Bannock War had broken out on nearby Camas Prairie induced them to retire to Challis. After the Indian hostilities were over, Smiley returned in October with T.B. Mulkey to locate a number of lode claims. Being an experienced Montana prospector and Utah mill superintendent, Smiley had no trouble raising another party as early in 1879 as they could get back into the high country. E.M. Wilson had greater success than any of the others by discovering the Vienna lode on June 4. A gold rush in the summer led to additional discoveries on Beaver and Lake creeks. A mining district was organized, and Sawtooth City was established on Beaver Creek. Rich silver ore, augmented by gold, was packed from the Pilgrim mine at Sawtooth to Atlanta that fall. San Francisco investors purchased the Pilgrim for $30,000 in the fall of 1879 and soon spent $45,000 more developing the property. Investors from LaCrosse and Winona, Wisconsin, took over the Vienna mine, so that during an enterprising initial season, Vienna and Sawtooth City moved rapidly toward production.

Developed between a gold rush to Yankee Fork in 1878 and 1879 and a silver rush to Wood River in 1880, Vienna and Sawtooth City enjoyed a considerable boom of their own. Attracting experienced miners from Rocky Bar, Atlanta, and Bonanza, these camps underwent extensive early exploration. With a 1,200-foot tunnel, a 100-foot shaft, and some raises which reached the surface, the Pilgrim lode (with a width from 10 to 25 feet) contained high-grade ore with assays ranging from 3,000 to 5,000 ounces of silver per ton. An average of 60 ounces a ton for 10,000 tons developed by

1882, and 2,000 tons more already for milling, justified considerable effort to provide a recovery process suitable for these sulfide ores. Two adjacent mines, developed by tunnels 250 and 360 feet long, had paid their own development costs, with several tons of high-grade ore running from $500 to $2,000 a ton sacked and ready for export. Yet another property, the Silver King, had 15 tons of $500 ore on hand by 1882, sorted into high-grade (4,400 ounces a ton), second grade (770 ounces), and mostly standard grade (93 ounces). A New York company had purchased two other prospects, Columbia and Beaver, in 1880 for $12,000 and engaged in substantial additional development. With construction of a toll road from Ketchum, two mills were brought in and that isolated camp, Sawtooth City, flourished during an expensive development phase.

The expenditure of $100,000 on development preceded any effort at milling. This wise precaution would have paid off better if a short freighting season had not delayed the completion of milling facilities until 1882. After being caught at Galena Summit by an unexpected early winter snow, freighters could not get their mill on to Sawtooth until midsummer. After all the delay, only a test run, capable of demonstrating an efficient leaching recovery process, was managed in August 1882.

Sawtooth City had grown into a substantial community to accommodate all the activity: three saloons, a general store, a meat market, two restaurants, a Chinese laundry, a blacksmith shop, and an assay office served most needs of eighty or ninety construction workers who had a sawmill to provide lumber necessary for erecting milling facilities. With a strong force of miners as well, Sawtooth City gained importance as a mineral center with lots of activity, but almost no production, to its credit. Even though short seasons and exceptionally difficult transportation interposed severe barriers to mining at such a remote location, proper development prior to mill installation showed that Sawtooth City was avoiding errors which had disrupted earlier mining at nearby Atlanta and Rocky Bar.

While miners at Sawtooth City struggled to get from development into production, adjacent Vienna had undergone substantial growth as well. Three general stores, six restaurants, two meat markets, a bank, a hotel, two livery stables, and fourteen saloons provided a good index of Vienna's prosperity. Like Sawtooth, Vienna had a sawmill. From July 4, 1882, until consolidation with the *Ketchum Keystone* in November, the *Vienna Reporter* provided news coverage of this interesting camp.

The development of the Vienna mine, while less

Vienna (c. 1884)

extensive than all of Sawtooth's properties, had justified construction of a major mill. A 7-foot vein of $200 ore (but with assays as high as 19,000 ounces of silver a ton running far above this average) was exposed by two upper tunnels each of which ran 275 feet to reach the Vienna lode. A tunnel was driven to provide access at a depth of 500 feet below the Vienna outcrop. While all this exploration was under way between 1880 and 1882, a twenty-stamp mill was completed in 1882. Built to precisely the same specifications as the highly productive Custer mill on Yankee Fork and installed by the same contractor, this plant exceeded $200,000 in cost. Production had to be delayed until the next season for lack of mill supplies when winter snow isolated Vienna and Sawtooth for another season.

A number of other mines near Vienna showed good promise as well. Two tunnels, one a hundred feet above the other, had produced 1,700 tons of 75-to 100-ounce ore by 1882 in the Mountain King mine, whereas another property had a blind tunnel that accidentally produced 30 tons of high-grade ore which sold for $10,000 prior to processing. Another 3- to 4-foot vein, parallel to the Vienna lode, was crosscut by a 160-foot tunnel from which a 75-foot drift developed high-grade ore ranging from $150 to $450 a ton. Another group of six claims, purchased for $50,000 in 1881, had 60 to 200 ounces of ore per ton revealed by two tunnels of 250 and 450 feet with a connecting 215-foot winze. Somehow many of these mines, which had such great promise in 1882, could not get into major production. But the Vienna lode began to average $750 a day the next season, returning a profit of $17.50 a ton on ore processed at $20. Activity reached a peak in 1884, with a cumulative production of $500,000.

Lake Canyon, west of Sawtooth City, also had rich lodes. Lacking their own mill, these mines provided high-grade ore worth shipping to Atlanta for processing in the Buffalo mill. A New York investment of $50,000 in 1881 led to considerable activity by 1883. Nine tons of ore that year returned more than $15,000 in Atlanta. Until the Vienna mine was shut down after 1886 so that milling of Lake Canyon ore could be transferred there, Lake Canyon was so isolated as to have little chance to recover much of its initial capital investment.

After establishing a mill that could operate successfully at Sawtooth City in 1882, investors could not match Vienna's success in mineral production. Unable to resume milling the next season until September 1, because of the "gross mismanagement of an ignorant and inexperienced man," they realized an encouraging return of $60,000 in custom milling of ores from other local mines. They needed to process their own ore as well, but faced a two-year delay trying to solve technological problems.

While a satisfactory recovery process was being sought, Sawtooth City miners had to ship high-grade ore to Ketchum's Philadelphia smelter in 1884. This necessity encouraged them to install a concentrator to process their stockpile of ore, which had accumulated over several years by then, to ship out with the high-grade. With concentrates running at $500 to $800 a ton, they finally managed to smelt much of the refractory ore which had resisted local milling. At last in 1886, they had solved the technological problems so that the mill could run all season. Two hundred miners finally were able to work at Sawtooth City, which belatedly joined Vienna as a successful mining camp.

But neither Sawtooth nor Vienna managed to continue its new found prosperity after 1886. William Hyndman managed to keep his Sawtooth City property going with eight to ten miners at work on $800 ore, but Vienna had only a watchman employed. Even Hyndman's operation had to shut down early in 1888 because his mine was accidentally flooded. Hyndman took over another Sawtooth property in 1888, but his success was limited to a twelve-ton shipment to Portland which provided a profit of $2,738.88 that July. A $60,000 development tunnel at Vienna, driven clear through to the South Boise face of the ridge, failed to produce any ore in 1888, so neither camp had more than limited operations until 1892. Lake Canyon ores gave the Vienna mill some custom business after 1886, but only a modest amount of high-grade could be shipped that far with any profit.

By the summer of 1892, William Hyndman had developed enough ore in his Sawtooth City property to keep his ten-man crew busy for a year. An expenditure of $25,000 for a new pumping system allowed him to resume full-scale operations. Then a disastrous shaft fire on August 9 wiped out his hoisting, pumping, and compressed air equipment. Unable to persuade his company to start over with new equipment, he had to shut down altogether. Neither Sawtooth nor Vienna produced much after that.

Later efforts to revive Vienna accomplished little. After Vienna's mines were sold at sheriff's auction in 1906, leasors did a little work in 1912. When they failed to engage in serious production, Vienna collapsed. Nothing but piles of lumber remained in 1914 from more than two hundred buildings that represented Vienna's early promise for success. Some ore was shipped out the next year, and a new camp and mill were erected at Vienna in 1917. All that effort went to waste because of failure to process any ore. Vienna wound up with an unused mill, although occasional leasors tried to resume production. As a result of extreme isolation and technological problems in ore recovery, Vienna never got much beyond a half million dollars in production by 1888, and Sawtooth City accounted for only about half that much.

Part III. Lead-Silver Development, 1879-1884

INTRODUCTION

Although Wood River lead-silver prospects had been noticed as early as 1864, their development came only after appropriate technology had been worked out in Nevada and Colorado. By 1878, lead-silver production in Eureka, Nevada, and Leadville, Colorado, encouraged Idaho miners to expand from gold and silver to base metal mining at Bay Horse. A rush to Wood River followed in 1880.

Within a couple of years, a number of other important lead-silver areas gained attention, along with a major copper district on Lost River.

Additional gold discoveries around Shoup maintained traditional mining interest during southern Idaho's lead-silver excitement. Others followed after Idaho's major lead-silver operations commenced much farther north in the Coeur d'Alene country in 1884. From then on, gold mining did not begin to compare with lead-silver recovery in

Idaho. Wood River became southern Idaho's leading mining empire. In later years, no other mining enterprises could begin to match Coeur d'Alene lead-silver production, which accounts for more than eighty percent of Idaho's total metal yield. Yet traditional precious mineral lodes increased their output during an era when lead-silver properties eclipsed them in overall importance. Gold and silver mining, which had maintained Idaho's mineral industry during hard and difficult times after the gold rushes, finally was helped by a still greater lead-silver development that emerged after 1880. Each of these transitions had momentous consequences in Idaho's mining history. After more than two decades of effort marked by uneven results and many failures, Idaho emerged with a large-scale corporate lode mining industry in which highly capitalized ventures displaced small or modest operations of earlier placer miners or lode promoters. New trends accompanied this change in which industrial labor had difficulties with mine management, and in which both faced problems accommodating to transportation and other commercial and economic variables beyond their control. Wood River mining development came in a context quite different from conditions faced by Idaho's earlier miners.

WOOD RIVER

Mineral discoveries on Wood River go back to the gold rush years following the Boise Basin mining excitement of 1862. A premature stampede to Wood River was reported early the next spring, and serious prospecting continued there in 1864. Nothing of great interest turned up then, except for Warren P. Callahan's discovery of a galena lode which he found along the Goodale's cutoff south of later Bellevue, when he was passing through the valley on his way to Montana. Prospectors set out for Wood River from Rocky Bar again in 1865. Most finished their search entirely disappointed, but a ten-man party found quartz veins interesting enough that they located two claims on September 11 near the divide between Camas Creek and Wood River in a district later known as the Hailey Gold Belt.

Indian opposition in part held back any development for fourteen years. Eventually two of the original discoverers returned during the Wood River rush and saw their mines flourish. Meanwhile, Warren P. Callahan came back to examine his galena lode near Goodale's cutoff. He and his brother located a claim on September 3, 1873, and followed up with another in 1874. They did their

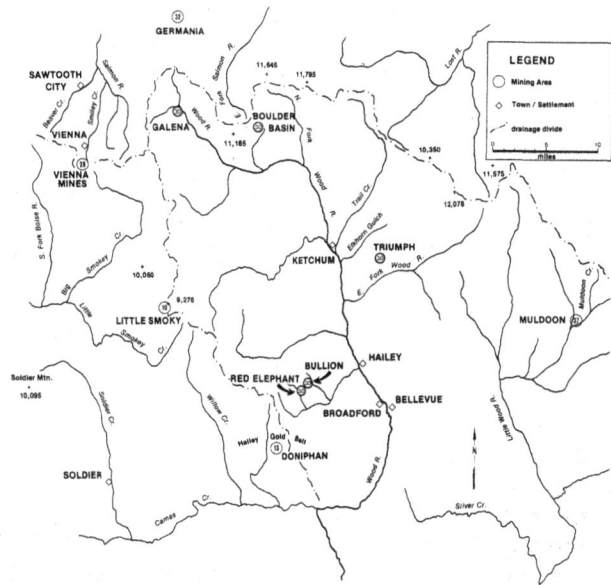

Wood River mining region. Numbers in circles refer to index map of mining areas, page 2.

annual assessment work regularly for the next four years. Then the Bannock War of 1878 kept them out. During that time, they displayed galena samples from their lode in Rocky Bar with little effect. Miners in Rocky Bar got excited about gold and silver, but disregarded lead.

By the end of the Bannock War, conditions had changed. Profitable methods of smelting lead and silver ores had been worked out in Eureka, Nevada, and improved still more in Leadville, Colorado, in areas where the ore could not be worked as advantageously as in Eureka. Better transportation was also on the way with railway construction extending through Idaho not far from Wood River. Lead prospects that amounted to little before the Bannock War now seemed worth developing. Warren P. Callahan came back to relocate his galena lode on April 26, 1879, and other prospectors began to look over the entire Wood River area. By the spring of 1880, the rush to Wood River began to transform the mining development of central Idaho, and some of the hopes of the early prospectors on Wood River were finally realized.

David Ketchum came across some promising lead-silver mines at the head of Wood River in May. Frank W. Jacobs found the Queen of the Hills, one of the major producers of the region, near later Bellevue (five or six miles north of Callahan's property) on July 15. Still other mines were found that summer near later Ketchum and Hailey. By this time W.H. Brodhead had many uniformly good assays (100 to 140 ounces of silver per ton) from the upper Wood River mines, where the town of

Hauling a boiler to Wood River

Galena was organized early in September. Jacobs City (renamed Broadford in 1880) followed almost immediately on lower Wood River where Frank Jacobs had turned up the Queen of the Hills. That fall W.P. Callahan shipped out a batch of ore to test in Salt Lake City: with an average recovery per ton of $431.46 in gold and silver, he proved that the new mines could pay well. Forty hardy settlers prepared to spend the winter in scattered parts of the extensive new mining region.

Thousands of fortune hunters joined the rush to Wood River in 1880. New towns, destined to over-shadow Galena and Broadford, sprang up in the valley. A post office called Ketchum was established on April 19, and the townsite for the new com-munity was laid out on May 2. (The townsite lo-cators still were calling the place Leadville, unaware that a postal clerk in Washington had refused to allow any more Leadvilles and had decided to call the place Ketchum, which he named for David Ketchum who had discovered the upper Wood River mines the year before.) Bullion City followed the discovery of the Bullion mine on May 28 and got a big boost with the addition of the Mayflower and the Jay Gould on July 4. Bellevue got a post office on June 23, and settlement followed quickly. With the discovery of the Minnie Moore, the most important of the early Wood River mines, at nearby Broadford on September 22, Bellevue showed genuine promise of becoming the major city of the region. A winter population of three hundred to four hundred stayed after the rush. Then John Hailey (whose Utah, Idaho, and Oregon Stage Company served Wood River until the railroad came) took up land between Bellevue and Ketchum on December 6, 1880, and that month another trading center was started on Hailey's site. Rivalry between Hailey and Bellevue

entertained the inhabitants of Wood River for the next several years. Hailey had the advantage of organized promotion by the Idaho-Oregon Land Improvement Company (a townsite company that also established Caldwell and Mountain Home and tried to take over Weiser) with an energetic developer backed by lots of capital resources. Robert E. Strahorn managed the promotion, and Andrew W. Mellon (later Secretary of the Treasury) got some useful business experience while a young man as the treasurer of the townsite company.

Shipments of ore from the Wood River mines were still fairly limited in 1880. Some Boise owners of the Idaho mine at Bullion sent out $17,000 in the fall, but that amount was only a modest beginning for what they had in sight. The other mines, even at Bullion, were not that far along. Smelters were needed close to the mines, and another Boise company, headed by David Falk and Alonzo Wolters, put up the Wood River Smelting Company plant at Hailey the next season. Hailey profited substantially from the early development of the mines at Bullion, which used the new smelter, and production rose greatly in 1881. With $80,000 to its credit, the Hailey smelter accounted for not much more than ten percent of the total. Another Wood River promoter, with interests around Ketchum in 1880, went east that winter to Philadelphia where major capital investment for the region was ob-tained. During the summer a large smelter was built at Ketchum, and other smaller ones were started for other mines. By this time lead miners at Eureka, Nevada, had learned that small smelters did not work very well, and the Philadelphia smelter profited from that lesson. It always had more capacity than it needed, and used the most advanced methods and equipment. Opening October 8, 1881, for a ten-day test, the Philadelphia

Opposite views of Main Street in Bellevue

Henry E. Miller built this mansion near Bellevue with his 1884 mining fortune (photograph taken in 1977)

PHILADELPHIA SMELTERS, NEAR KETCHUM, ALTURAS CO. IDAHO, PROPERTY OF THE PHILADELPHIA MINING & SMELTING CO.

Philadelphia smelter, Ketchum

smelter prepared for major production in 1882. The Wood River mines exceeded $1 million that year, and close to a fifth of the total was handled by the Philadelphia smelter.

Most ore was still being shipped out to Omaha, Salt Lake City, Kansas City, or Denver. At this point the Philadelphia Mining and Smelting Company decided to offer prices competitive with the outside smelters. The plant was doubled in size the next spring, when Idaho's earliest electric light system was installed. The Philadelphia company acquired new mines in 1882 to supply the larger smelting capacity, and other investors put considerable capital into similar mine purchases. Altogether, over $1.5 million was put into Wood River mines in 1882. Philadelphia and Salt Lake City were important sources for funds. Fourteen major sales took place in 1882, the largest being the Mayflower at Bullion, which was purchased for $375,000. E.A. Wall, who already had important

properties at Bullion, added the Bullion mine to his holdings, spending $200,000 for this acquisition. Sparing the $200,000 did not make too much of a problem for him: he shipped out more than $668,000 worth of ore from November 1 to December 10, 1882. The Bullion mine had only produced $320,000 when he bought it—production that amounted to little more than development work. The same could be said for most of the Wood River mines up to 1882.

⚒   🐴   ⚒

Transportation improvements—particularly the completion of the Oregon Short Line to Hailey, May 7, 1883, and on to Ketchum, August 19, 1884—allowed the Wood River mines to reach their maximum early output. Up through 1882 for the mines around Hailey, and until most of 1884 for those at Ketchum and beyond, miners preferred to

"Bullion Mine", Bullion, Alturas Co. Idaho. Property of The Wood River Gold & Silver Mining Co.

"May Flower Mine", Bullion, Idaho, May Flower Consolidated Silver Mining Co. of Chicago.

hold back production until they could profit by shipping at reduced railroad rates. They expected to save $20 a ton over wagon freighting. Rail transportation also provided faster, more comfortable passenger service. Until the Oregon Short Line entered the area, daily stage lines from the Utah Northern at Blackfoot and the Central Pacific at Kelton served Hailey and Ketchum. Even after the railroad arrived, stages and freight wagons still had to take care of places like Bullion and Galena that were too high in the mountains to be reached by rail. By the beginning of 1884, a new toll road up Trail Creek from Ketchum to Lost River and Bay Horse provided a route that served as a worthy test

Bellevue's Palace Club saloon and patrons (1905)

Bellevue (1909-1910) had a modern school (above) and opera house (below) when mining was active there

for H.C. Lewis' huge ore wagons that are still preserved in Ketchum. With the railroad boom, Bellevue, Hailey, and Ketchum reached their peak. For a time, Bellevue had two daily newspapers (until the railroad went on to Hailey), and Hailey had three. (These were not the earliest daily newspapers in Idaho by any means, but at that time they were the only ones.) With Idaho's earliest phone service, as well as Idaho's original electric light installations, Wood River rated as the most progressive region in the territory.

Producing over $2 million a year for the next three seasons after the railroad brought added prosperity in 1884, mining on Wood River offered wonderful returns to those fortunate enough to own the right properties. In a single year, Isaac I. Lewis' Elkhorn mine near Ketchum yielded ore sold to the smelter for $161,841.72 for a cost of only $35,372.33: while this did not amount to a really big producer, a net profit of over $126,000 a year—a profit of almost eighty percent of the yield—gave its owner capital to invest in organizing the First National Bank of Ketchum in February 1884. The largest of the early producers, the Minnie Moore, was sold to a director of the Bank of England on February 25, 1884, for $450,000 with ore reserves of $675,000 on hand. With a declining price for silver, operating expenses exceeded profits somewhat, but before the British operation shut down, the mine had produced well over $1 million more than had been blocked out at the time of

Hailey's Main Street before 1888

Hailey in the spring of 1888

Winter scene in Hailey (1888-1889)

Hailey, July 2, 1889

An early Ketchum mine

purchase, and profits far more than repaid the initial investment. An even bigger sale (unaccompanied by such spectacular returns) of the Bullion mine for $1,050,000 at the same time—with $685,000 in cash, and the rest in shares—also held great promise with as much as $1-$2 million worth of ore still in sight. This investment, also of British capital, marked the height of the early Wood River mines. When the Triumph mine was discovered on the east fork early in June, the owners declined a $40,000 offer for an undeveloped, but fabulous looking, prospect. Eventually the Triumph turned out more value than the $20 million from all the early Wood River properties combined—but it took until 1927 to get major production going. Some of the Triumph metal sold at higher prices, and altogether $28-$29 million was realized between 1936 and 1957. Long before that, the early Wood River mines had gone into decline.

⚒   🐴   ⚒

Labor difficulties, generally brought about by attempts to reduce miners' wages in the face of expected rising costs and declining prices, foreshadowed the end of Wood River's early prosperity.

On July 20, 1884, the miners at the Minnie Moore struck in protest for not being paid, and ten days later they won a settlement that warded off wage reductions which had been threatened. Early in 1885 the miners' union emerged less successfully from a similar dispute. Facing prospects of military intervention, the mine union lost its fight to maintain wage levels. After much protest and excitement around Bellevue, Miners went on to attain their peak production of the early period of operation. Cost reductions through twelve percent wage cuts helped the various Wood River districts to maintain high production for two more years.

An abrupt drop in output in 1888 (in which the total fell almost in half, but still exceeded $1 million), followed by a much more severe collapse after 1892, reflected a declining price of silver. What had been low-grade ores were ruined, and what production there was had to be shipped out for smelting. The Philadelphia smelter had to shut down with the 1888 decline: with capacity to handle everything that Wood River produced, the Ketchum smelter was technologically handicapped by being in a remote location in which trained specialists were hard to find and where repairs on equipment were hard to make. Large smelters in places like Omaha got ore from many different places and could mix various ores in the combinations needed to satisfy the complex chemical requirements for processing refractory lead ores. In the early years of high-grade production the central Idaho smelters could afford to ship ore back and forth by wagon to each other to

SHEEP MOUNTAIN, GREYHOUND RIDGE, AND SEAFOAM

Miners' Union parade in Ketchum, July 4, 1884

meet such needs. The Ketchum smelter could afford to import iron ore from Wyoming and coke from Utah and to make large amounts of expensive, and not too satisfactory, charcoal in twenty-one charcoal kilns. After the Philadelphia smelter in Ketchum shut down, costs of sending ore by rail to distant smelters ($10 or more a ton) exceeded the actual smelting cost of $6.50 to $7.50 a ton. Reopened from 1902 to 1906, the Minnie Moore put out more than another $1 million in spite of such costs. But by then, the early years of mining prosperity on Wood River were over.

Later the Triumph, with a yield of about $28 or $29 million between 1936 and 1957, more than matched the early production. Rising silver prices in 1967 led to revival of some of the old lead-silver-zinc properties around Bellevue, with production of $1,574,000 in 1967, and nearly $2,000,000 in 1968. When this operation shut down in April, 1970, Wood River finally had reached a total of over $62 million.

Recent Wood River mining develoment included a five-hundred-foot tunnel driven in 1977 on a vein discovered near Ketchum in a road cut. In addition, reprocessing of Minnie Moore dumps near Bellevue brought renewed activity at that early producer.

While out looking for Indians during the Sheep-eater campaign in 1879, Colonel Reuben F. Bernard found an interesting lode prospect on Sheep Mountain on June 8. When the army returned that way on August 31, Manuel Fontez and his packers hauled a number of samples back to Boise. Obtaining good assays from his test specimens, Fontez set out with a small prospecting party in the spring of 1880. They found several good leads on Greyhound Ridge and created enough interest that several hundred miners showed up to prospect there each summer after that. John Early had particular success locating galena with Fontez on Greyhound Ridge the next summer, and Fontez found attractive outcrops on Sheep Mountain in 1882. Within another two or three seasons, about forty-five claims had been recorded in these adjacent mining camps. Jesus Urquides, Boise's most prominent packer who had come out with Fontez and taken up a productive Greyhound claim with John Danskin, spent the summer of 1885 hauling ore out to a smelter at Clayton. Plans to extend a road from Cape Horn to Greyhound Ridge and Sheep Mountain were contemplated as a means of developing the isolated prospects, retarded as they were by their location remote from transportation in rough country high above the Middle Fork of the Salmon River.

By 1886, additional discoveries along the Cape Horn-Loon Creek trail, a dozen miles southwest of Sheep Mountain, had extended even more exciting mining prospects to Seafoam. That summer, a series of scattered new lodes offered considerable hope of major production. A twenty-foot tunnel in the Summit mine at Seafoam exposed a twelve- to fourteen-inch vein of ore running as high as $1,000 a ton. The samples of sulfide and chloride ores had to be smelted at Clayton, where reasonably close facilities made testing convenient. Another Seafoam property had samples processed in the Custer mill with encouraging results ranging from 100 to 600 ounces of silver a ton. Still another pack train lode went for $325 at Clayton's smelter. Discovery of the Josephus mine at Seafoam in September 1886 offered still more hope. An "immense body of float," distributed along a lode pattern 1,000 feet long and 20 feet wide, substituted for an outcrop. Assays ranged from 400 to 700 ounces of silver a ton. Before silver prices fell drastically in 1888, these properties induced substantial investment and activity.

With still more dramatic lead and silver price

Greyhound mill and smelter (1909)

declines in 1892 and a national economic panic the next year, Seafoam, Greyhound Ridge, and Sheep Mountain failed to prosper. Extensive investment from Salt Lake City brought serious development there toward the end of the century. When A.C. Bomar visited the area late in 1899, he reported to the *Salt Lake Tribune* on the best developed mine at Seafoam:

> We saw over 1000 tons of ore on the various dumps, some running up into the hundreds of ounces in silver and as high as $20 in gold.
>
> An open cut is made on the lode twenty feet wide, with the ore assuming better value every foot driven.
>
> Another cross-cut tunnel was being driven 800 feet east of this which, when completed, will be 350 feet long and will cut the vein 350 feet deep. At the entrance of this work is a natural millsite and an inexhaustible amount of splendid timber. It is estimated that between these tunnels (800 feet) there will be opened up $5,000,000 to $7,000,000 worth of ore, which with the facilities offered can be worked (mined and concentrated) for $1 per ton. Various open cuts and shafts are made the entire length of the claims (6000 feet), all of which show ore that will be profitable to handle.
>
> Four lakes touch the side lines of the group

and two small streams cross them, insuring ample water for power and domestic purposes. It is estimated that within 100 feet of the surface of this group there is 1,000,000 tons of ore, worth $20,000,000.

Another Seafoam property had six hundred sacks of high-grade ore ready to be packed out as soon as possible in 1900. With assays as high as $10,000, $15,000, and even nearly $25,000 a ton (but only for small samples), and with other "fabulously rich ore" nearby, Seafoam held great attraction. Rail transportation was needed to ensure that all this wealth could be recovered. The construction of the Idaho Midland—a Boise line headed for Butte—commenced at Boise on May 8, 1900. Once the line reached Cape Horn, Seafoam would no longer undergo the isolation that had retarded mineral development. Somehow the Idaho Midland never got started. Extending rail service from Mackay to Challis might have helped, but that project failed also. A decade later, after the Gilmore and Pittsburgh reached Salmon, an expansion of the rail system up Salmon River was contemplated.

Extensive prospecting directly west of Greyhound Ridge and Seafoam, incidental to the

Dog sled leaving from Stanley for Seafoam, led by "Speed"

Thunder Mountain gold rush of 1902, broadened the area's interesting lode possibilities. J.W. Speeks, who had discovered a Middle Fork lode just below Soldier Creek in 1894, shoveled snow on trails to Thunder Mountain in 1902 in order to return before someone else could take over his prospect. He located four promising veins on his second attempt. None became productive in that rough country, although one twenty-foot ledge carried assays of $20 a ton in gold, augmented with twelve ounces of silver. Later that summer, W.B. Patten, who had operated mines in Colorado, Utah, Idaho, and Montana, spent ten weeks exploring a lode on Indian Creek. When he returned to Boise on October 5, with almost fifty pounds of a variety of mineral samples including wire gold, he set off still more excitement. His efforts to introduce Denver and Pueblo capital failed to produce important Indian Creek mines. Only some very nominal returns came from Middle Fork placer efforts below Greyhound Ridge and Seafoam.

Major investment returned to this area in 1926 when Hecla (a major Coeur d'Alene mining corporation) acquired mines at Sheep Mountain and Seafoam. Seventy miners at Josephus Lake deepened a shaft there to 250 feet before work was suspended. More development work resumed a year later. A sawmill, a 230-horsepower hydroelectric plant (with a 4½-mile power line), a 350-cubic-foot compressor, 3,400 feet of flume, six houses and two large bunkhouses, and a 50-ton mill were constructed. Yet major transportation improvements did not arise from this activity, so Seafoam remained isolated.

Aside from severe transportation difficulties, recovery problems hampered the development of Seafoam's grand potential as a mineral empire. Even though limited production came from additional prospects on Pistol Creek Ridge west across

the Middle Fork of the Salmon River from Greyhound Ridge, the area disappointed a substantial group of enthusiastic investors. A small smelter was finally installed on Greyhound Ridge, but less than $400,000 out of an anticipated $20 million came out of Seafoam, Sheep Mountain, Greyhound Ridge, and Pistol Creek Ridge. Out of this unfortunate experience, mining engineers learned to distrust rich lead veins in granite formations.

Production at Greyhound resumed in 1979 after another mill was installed, and activity continued there in 1980. Exploration and testing of Sheep Mountain properties also was undertaken in 1980.

GERMANIA BASIN AND EAST FORK OF THE SALMON

Galena discoveries in Germania Basin expanded mining prospects north from Wood River to the upper East Fork of the Salmon River in the summer of 1879. Two Germania lodes about five miles apart had promising assays during the initial season. Samples of silver running four hundred ounces per ton came from a rich fourteen-inch seam in a three-foot vein in one, and values as high as $240 came from another ten- to twelve-foot vein that generally ran $12 in gold with $30 in silver. James Steele and William Short exposed six hundred feet from the ten- to twelve-foot vein that summer. By October, packers had hauled some of the better ore out to the Utah Northern at Blackfoot for shipment to a smelter.

Work at Germania continued steadily, and 150 tons was packed to the Clayton smelter in 1881. In the spring of 1882 five men got 50 to 60 more tons of high-grade ore ready for shipment from Fred Spereling's Germania mine to a new smelter only eight miles away at Galena on upper Wood River.

That summer a daily production of 10 tons of ore assaying 150 ounces of silver per ton was stockpiled while a road was under construction to accommodate teamsters instead of packers. By the next summer, development of the Livingstone mine on Railroad Ridge exposed assays of 100 to 400 ounces of silver per ton in another east fork property only twelve miles from a convenient smelter at Clayton. In 1890, an eastern company commenced to purchase the Livingstone group with a $100,000 intial payment. By 1902 S.A. Livingstone was working this property with some success, although major production was deferred until 1926.

Southern Idaho's major lead-silver-zinc discovery in thirty years revived mining at Livingstone in 1925. A new plant and equipment, installed in 1926, made this southern Idaho's largest mining employer and next to the largest metal producer within another year. Active from 1926 to 1930, the Livingstone accounted for $650,000. Prospecting for molybdenum led to major discoveries near the Livingstone close to Castle Peak in 1968 and 1969, when the report of a $100 million orebody attracted wide attention. Creation of the Sawtooth National Recreation Area in 1972 retarded the development of this politically controversial mining claim, but American Smelting and Refining Company's interest there led to renewed effort in 1980 and 1981. Unfavorable prices, however, deferred additional development until after 1982.

ERA AND MARTIN

Not far from Goodale's cutoff—an Oregon Trail emigrant road which ran near Arco, Carey, and Fairfield—lead-silver lodes brought miners to an area close to Craters of the Moon. James B. Hood, who participated in making the mineral discoveries there in 1879, spent several years convincing potential investors that he had anything of value. After the Wood River rush of 1880, he had less trouble. A Blackfoot-Wood River stage and freight road ran along Goodale's route, making Hood's potential mining camp more accessible.

Two years after he began working, Hood had eight lead-silver veins ready for development. One had a twenty-foot shaft which exposed a lode that tested as high as one hundred ounces of silver per ton, enriched slightly with $2 more in gold. Outcrops from other veins assayed from one hundred to 200 hundred ounces, and two camps began to grow up there. James Hood's location came to be known as Era. Frank Martin's valuable Horn Silver mine supported Era even more than

Hood's property did. About four miles away another, longer lasting community named for S.D. Martin (Frank's brother) flourished for a time.

Development proceeded slowly at Era. Hood drove a 240-foot tunnel in 1882 and began to build an ore reserve. By the middle of 1884, two loads of Era ore reached Hailey for testing. These yielded 485 and 633 ounces of silver a ton. Then Frank Martin's 4-ton shipment to Salt Lake City produced 814 ounces a ton. Eastern capital was attracted by these encouraging values. Frank Martin, whose Horn Silver ore did so well in Salt Lake City, decided to employ twelve miners to develop his property. On August 1, 1885, he managed to sell his Horn Silver mine for $62,500. Era began to thrive. A townsite had been established for Era in the spring of 1885, and another townsite at Martin followed late that year. (Martin, in fact, had a post office from June 21, 1882, to April 30, 1940, while Era had one from August 26, 1885, to July 5, 1894.) Additional discoveries eight miles away on Antelope Creek in May 1885 soon created still more interest in that promising area. By August, Era had a dozen tents, four or five cabins, and a few frame buildings under construction. A general store, a boarding house, a restaurant, a blacksmith shop, a barber shop, a hotel, and three saloons occupied those tents and cabins. All these services were available to potential miners who, as yet, had no possibility of employment. Forty people spent the winter in hopes that operations eventually would commence.

In 1886, investors from Salt Lake City provided Era with an economic base that had been needed badly for a year or more. A twenty-stamp mill, capable of expansion to forty stamps, was built that summer. Around eighty freight teams, many provided by Mormons based in a camp of their own two miles away, were employed to haul in 750 tons of mining equipment. With $100,000 invested in a five-level mill and recovery plant that utilized a roasting process, and another $100,000 devoted to development of the Horn Silver mine, Salt Lake City capitalists installed a fine modern facility. Their mill had a handsome electric light system a year before Boise had electric power, and by the fall of 1886, twenty-five miners and another twenty-five mill workers were employed. For a while that winter they produced $7,000 a week. The next summer, they managed to ship $1,600 to $1,800 on alternate days. Forty-eight miners worked underground by that time. About $250,000 was realized in 1886-87 from their Horn Silver mine. Generally, though, their milling experiment failed. Sometimes they managed to operate as a custom mill, but by 1888, about $5,000 worth of Era ore from a major new 1887 discovery was hauled to Nicholia for

Era in 1888

smelting. Then in 1888 a price collapse created additional problems. After Nicholia's smelter shut down, Era had an inactive, slightly used stamp mill and no good processing plant capable of handling local ore available within a reasonable distance.

In spite of recovery problems, J.W. Ballentine managed to sell his galena mine at Era — the one which had shipped to Nicholia — for $90,000 to Alexander Majors of Kansas City, who was joined by a group of Wyoming investors in the summer of 1889. They employed thirty-five or forty men to bring in new machinery and to sink a 275-foot shaft from their 175-foot tunnel on the promising new lode. They leased Era's existing twenty-stamp mill but could not perfect a recovery process either. Although they tried twenty-five percent wage cuts (an expedient that lost them only a fifth of their employees), they still could not operate profitably. By November 1889, their operation was shut down by judicial injunction because they failed to pay $40,000 of the purchase price. Leasors on Martin's Horn Silver mine kept up a little activity at Era, but litigation kept the other major property inactive for two years. Finally one of the Ketchum sellers bought the property back at sheriff's auction on November 7, 1891, for the $40,000 overdue payment. But little else resulted from that transaction.

Milling resumed at Era on July 1, 1893, with a force of seventy miners, but by fall, this operation had failed also. An end of costly litigation in 1894

helped revive the major mine there. Some modest small-scale activity went on until 1897, when major development resumed. Additional efforts in 1901 encouraged the camp. A small amount of lead was recovered in 1908, and subsequent brief revivals came in 1913 and 1928. In spite of repeated obstacles and recovery problems, the region finally had managed to produce about $400,000 by 1900.

MACKAY AND COPPER BASIN

More than three decades of notably unsuccessful effort went into opening Idaho's major copper lode at Mackay. Not as isolated from transportation as Levi Allen's Seven Devils discoveries, large copper properties near Mackay had to wait for rail transportation to realize their potential. Unlike many rich Idaho mining areas, Mackay could be reached without excessive difficulty. Yet until a mining company with resources sufficient to build a rail line to Blackfoot developed enough ore to warrant investing in supplying such service, mining at Mackay could not get under way on any reasonable scale.

Prospectors who found copper possibilities along Alder Creek in 1879 progressed very slowly in trying to exploit their potential mines. Too poor to be able to spend much time trying to figure out what they had, they accomplished little toward preliminary exploration of the new mining area for four years. Finally additional discoveries set off a boom in 1884. A group of new camps suddenly sprang up. Some emerged as ghost towns after only a year or two; others went through occasional phases of activity between longer periods of collapse. Houston, Alder City, and Carbonate all got under way early in 1884, and Cliff City followed that summer. No more than two of these initial four really were needed, so Alder City and Carbonate promptly failed. Mart Houston got a post office established on January 25, 1884, at a superior location, and that spring a gin mill followed. Houston had a half dozen houses (twice as many as Alder) by June, but Carbonate had only three partly completed dwellings. Alder City gained a post office on April 22 to offer Houston some competition, and soon had five businesses and a saw mill. Houston came out ahead, however. By early summer, twenty more houses were under construction, with stores and "a commodious lodging house" following by August.

Chosen as a site for a copper smelter, Cliff City, four miles from Houston, flourished until construction was completed that fall. A store, twenty houses, and two or three saloons augmented Cliff City's twenty-ton smelter. Houston grew even faster, favored by a location along a route to Cliff City. Alder City's remaining store and whiskey shop disappeared, the log structures being moved to Houston. Carbonate's "solitary gin mill" also was deserted.

Smelting commenced November 23, with a pure looking product recovered from an initial batch on November 24. Early in December, after a test of little more than a week, Ralf J. Bledsoe shut down his smelter, discharged his entire crew, and left Houston, a town now a little better off than Alder City and Carbonate. A single mine employed twelve men the following spring, while farms and ranches began to flourish. When smelting resumed at Cliff City on July 22, 1885, Houston had a short revival. By January 1886, Cliff City's smelter had become idle again. Until technological problems could be solved, Houston and Cliff City faced another depression. Houston had become a family town, with a school of sixty pupils. Yet by June 1886, only eight inhabitants remained. Most of Houston's sixty-five buildings stood empty. A concentrator helped revive Houston in the fall of 1887, although failure to equip the plant for winter operation forced the suspension of production until the spring of 1888. During 1888, however, attention was diverted to Copper Basin, west of Mackay, where assays of thirty-five to fifty-five percent copper attracted interest to a promising new lode. Silver values raised Copper Basin ores to $145 a ton in a thirty-five-foot vein. After several years, preparations were made to haul some of this ore down Trail Creek to Ketchum for shipment to a Salt Lake City smelter.

New York investors finally came to Houston's rescue in August of 1890. A young and inexperienced mining superintendent spent $100,000 in twenty-six days, all to no advantage. Yet vacant houses were reoccupied, and new ones were constructed. John Danskin contracted to deliver thirty tons of ore each day to Cliff City's smelter, which ran from late in 1890 through February. Then residents of Houston and Cliff City became despondent when mining operations suspended.

Another revival stirred up hope in Houston in June 1892. British investors financed the construction of a good road to a large Alder Creek mine. Development work had barely begun, however, when "the eccentric Englishman in charge" abruptly returned home to Sheffield in August 1892. This turned out to be a poor time for mining expansion, so more delays ensued.

After several unsuccessful attempts to revive copper mining around Houston during the Panic of 1893, new capital was finally attracted there. W.A. Clark of Butte, whose copper holdings dominated that major mining area, undertook an ambitious program during the summer of 1894. He began shipping copper bullion from an old smelter and put an additional furnace into operation on September 16. Houston flourished again, although Clark was engaged only in test operations, which he described as prospecting. By 1899, a shaft at White Knob had been sunk to a depth of seven hundred feet, at which point an access tunnel was driven so that ore need not be hoisted all that distance. Many irregular orebodies, diverse in character, had been developed, and a recovery technology capable of handling this variety of ores was employed. Only $8,000 worth of copper was produced in 1899, but costs had become more favorable. Not much lumber could be found nearby, so $15 a thousand feet had to be spent for timber. Fuel wood ran $4 a cord. Freight for importing supplies costs $12 a ton, but ore could be shipped to a rail terminal for $10. Ore could be smelted for $10 a ton, so low-grade rock, available in great abundance, could not yet be processed. Miners worked eight to ten hours a day for $3 to $3.50 while engineers made $4. At that wage level, tunnels could be driven for $8 a foot, and shafts sunk for $45 a foot. Major investments were needed to cover these expenses.

Finally John W. Mackay of San Francisco (the most prominent of four miners who had developed Virginia City's major Comstock property) became

Empire mine at White Knob

interested in White Knob. He arranged with the Union Pacific to build an Oregon Short Line branch from Blackfoot to his new mine. At this point, Houston, which had grown to two stores, a restaurant, a boarding house, a blacksmith shop, two livery stables, four saloons, and a Methodist church, was replaced by a new rail terminal three miles away. There a new mining center — Mackay — gained a population of about one thousand two hundred when rail service reached Lost River in 1901.

Preliminary development of Mackay's ambitious operation encouraged the construction of a six hundred-ton smelter. A million tons of ore containing four percent copper, with about $3 more in gold and silver, had been located. Wayne Darlington ran a series of encouraging tests in a fifty-ton smelter to verify that different deposits in White Knob's primary lode all could be processed without insurmountable difficulty. His various test lots produced 200,000 pounds of copper by direct smelting. Five hundred men were employed building a new six hundred-ton smelter with a twelve-mile electric railway for transporting ore. Along with this ambitious project, Mackay's firm developed Copper Basin properties.

Long before Mackay's smelter was completed, mining at White Knob underwent important changes. Wayne Darlington had been unable to get along with his miners, who finally decided to organize on April 4, 1902, as a union in the Western Federation of Miners. (The Western Federation had enough strength at Bay Horse to win a strike to preserve wage levels and had powerful unions in other camps such as Custer and Gibbonsville.) Only two days after they organized, White Knob's miners had to strike in order to avoid being driven out. In this showdown, Darlington's services were dispensed with, and his "superintendent and foreman likewise left the country." Mackay then became a union

town, with a union restaurant. Prior to their strike, all miners had to use company facilities. "Married men were not allowed to board with their families, but now single men, as well as married men, can board where they please."

Darlington departed at a time when his six hundred-ton smelter was about half finished. Only about three months later, John W. Mackay died in London, July 20, 1902. When one of two projected furnaces was completed in October 1902, Darlington's direct smelting system (with no preliminary milling) recovered all but a half percent of high-grade copper that made up fifty-eight percent of the ore being processed. With enough ore developed to assure a $2 million profit, Mackay retained a permanent population of almost a thousand. A new crosscut tunnel, intended to reach the White Knob lode at a depth of 1,600 feet (900 feet below an existing 1,100-foot tunnel) was commenced. But a second furnace required to raise Mackay's daily smelting capacity to six hundred tons was abandoned.

⚒   🐴   ⚒

In 1904, mining operations at Mackay shifted to a new system. Ravenal Macbeth, whose Lucky Boy property at Custer had become unprofitable, took a White Knob lease. From that time on, most underground work was carried on by leasors. A year later, a matte process was employed in Mackay's smelter with unfortunate results. The lack of sufficient sulfur in White Knob ores accounted for technological difficulties. Costs also escalated after mining and smelting services had fallen into "the hands of a number of high-priced operators." Frank M. Leland was sent up from California to dissolve the company and to dismantle the matte plant.

Leland had an unusual gift for reducing costs drastically and for inventing economical processes. Disposing of "superfluous supplies and equipments" in order to obtain operating capital, he employed a competent assayer and resumed smelting a reserve of the low-grade ore on hand when his company had failed. Contracting with "some intelligent leasors" to supply enough high-grade ore to provide a mixture suitable for smelting, he ran his furnace for several months. Then he replaced an expensive electric railway with a Shay steam locomotive, cutting transportation costs to a fourth of their previous level. So instead of closing out White Knob, he commenced a system in which different leasors working in different parts of his lode supplied his company smelter. Within two years, he had "rescued and transformed [White Knob] from a dismal failure to a large and profitable producer of copper."

In 1907 the Empire Copper Company of New York assumed Ravenal Macbeth's lease. Instead of operating a three hundred-ton Mackay smelter at considerable expense—partly for importing Bingham Canyon ore from Salt Lake City, as Macbeth had to do in order to supply a sulfide deficiency—Leland sent all his ore to Salt Lake City for smelting. That way he got rid of shipping costs for hauling copper (in ore) up from Salt Lake City and then sending his copper back again. Great care had to be exercised in cost controls in order to process low-grade ore ranging from sixty to eighty pounds of copper a ton. A copper price collapse during the Panic of 1907 forced him to shut down on October 1. Unable to resume until copper prices rose to 17 or 18 cents a pound, he had to wait a year or two. But he had developed a leasing and smelting system that would work when a price recovery would make mining feasible again.

By 1910, leasors for Empire copper operations brought out ore sufficient to produce 830,000 pounds of metal in the Garfield smelter near Salt Lake City. A Copper Basin mine also shipped ore to Garfield, and in 1914 Lost River discoveries west of Copper Basin next to Mount Hyndman increased the copper zone still further. Mackay produced steadily until adverse prices greatly reduced White Knob's yield in 1914. Recovery in 1915 enabled leasors to employ two hundred men for a record annual production. In 1916 an Empire dividend of $250,000 rewarded investors after a decade of more modest return. That year, a lower tunnel, driven 6,000 feet to reach ore at a depth of 1,600 feet, was completed at last. An inexpensive tram replaced the Shay railroad in 1918, reducing transportation costs by eighty percent between White Knob and Mackay. Such savings had become more than necessary to compensate, in part, for high war-time operating costs which had increased with wage raises and rail car shortages. Copper Basin mines had profited by better prices after 1916, but faced problems comparable to White Knob's in 1918.

By 1919, low prices and high production costs restricted production at White Knob severely. Leasors shipped only 10,000 tons of ore containing five percent copper. After a long shutdown, production resumed in November 1921. Price increases accounted for this revival, which allowed leasors there to make White Knob into Idaho's leading copper producer for a number of years. Operations continued steadily until July 1, 1924, when fire destroyed an ore bin and tramway head house. "The tramway was also badly damaged by buckets that were liberated and ran wild during the fire." Conditions favorable to copper mining allowed rapid reconstruction so that operations could resume on September 25.

For nineteen years, leasors had been filling White Knob stopes with low-grade ore. Enormous reserves had accumulated by 1924. A 150-ton floatation concentrator was installed to process this previously unmarketable copper. Enlargement of this mill to 250 tons the next year provided work for a hundred miners on a leasing system in which their return varied with copper price fluctuations and the percentage of copper in the ore. This arrangement continued until 1928, when further development (that could not be provided for under such a lease arrangement) became necessary. A new corporation acquired the Empire property, rebuilt the plant, and engaged in ambitious development. This program continued until August 1930 when the early problems of the Depression forced a shutdown. Leasors again took over, although they could manage only to build up ore reserves, which they could not afford to ship to a smelter. Leasing continued after a tax sale in 1931, but Depression prices did not favor copper mining.

Before a copper price recession in 1914, White Knob had produced about $3.75 million in copper (copper made up two thirds of this total), silver, and a little gold. Later yields, supplemented with some tungsten production from nearby Wildhorse (1953-1955) raised this total to around $15 million. In reacting to a variable market economy with abrupt price fluctuations over many years, copper production at White Knob of necessity lost efficiency. Additional elements of production not readily subject to management control included problems and costs of transportation along with technological innovation essential to facilitate production from low-grade ores. If White Knob had benefited from a location like Bingham Canyon near Salt Lake City in Utah, where much lower grade copper ore could be processed to advantage, Idaho could have had a large copper output. (Much of Bingham Canyon's advantage came from the development of large-scale open-pit mining—a possibility inappropriate at White Knob.) Yet mines such as those around Mackay and Copper Basin deserve credit for enabling smelters such as Bingham Canyon's plant at Garfield, Utah, to provide for increased recovery of copper ores from other places. Smelting at Garfield depended upon having a great variety of ores to provide a mixture of minerals essential for efficient operation. White Knob contributed more than its share of different ores required to operate a smelter elsewhere. Credit deriving from this kind of incidental benefit has gone to other places. Yet White Knob deserves some recognition beyond a relatively modest copper production.

MINERAL CITY

During the presidential campaign of 1880, John A. James and Jim Peck crossed the Snake River from Baker to prospect south of Heath below Sturgill Peak. They found good silver lodes with scattered superlative assays comparable (as so often was the case) with those of the Comstock lode in Nevada. Naming their lode for presidential candidate W.S. Hancock, they had to give up after their new camp finally became known as Mineral City. They eventually figured out that they did not have another Comstock lode either. A mining district was organized there on September 22, 1881.

An early milling experiment with untested equipment failed and some expensive litigation over claims set their mine back as well. Extensive Salt Lake City testing of Mineral City's refractory ores led to installation of an 1889 smelter, financed substantially by J.W. Huston not long before he became chief justice of Idaho's Supreme Court. Another small smelter was tried in 1889, but both failed. Finally a new smelter installed in 1890 did better. Mineral City flourished until declining silver prices proved ruinous in 1893.

Aside from litigation and problems with smelting technology, Mineral City had some advantages. Located only four miles from a branch rail line down the Snake River, the community had little difficulty in obtaining supplies or in shipping ore. Slag from the smelter provided good local building material, and unlike many Idaho mining camps, Mineral City was not isolated by deep winter snow. Local fruit and gardens flourished.

An effort to revive Mineral City with a large smelter failed in 1900. Another smelter failed in 1902. Then European investment supported a camp of sixty to seventy miners and a smelter that achieved most of Mineral's production (a million ounces of silver) before a labor strike protesting wage reductions shut down the camp in July 1904. War-time revivals in 1918-1922 and for a few years after 1940 resulted from the higher prices during those periods of national emergency.

GILMORE AND THE VIOLA MINING REGION

Lead-silver mining areas of substantial magnitude made the Lemhi Range near Birch Creek divide an important mineral region for a half century after 1880. Together with the celebrated Viola mine across Birch Creek near the Continental Divide, the area accounted for more than $16 million of Idaho's metal production.

Interest in this area goes back as far as the mining rush from Leesburg to Little Lost River on June 15-16, 1867. For more than a decade after that, lead-silver lodes offered little attraction for prospectors. An upper Lemhi mining district was finally organized in the Gilmore area in 1880. Additional lead-silver discoveries eight to ten miles south on the same vein system brought prospectors to Spring Mountain. Discovery of the Viola mine expanded the mineral area across Birch Creek practically to the Continental Divide. In the fall of 1881, C.F. Blackburn came across a Birch Creek lode of galena and silver carbonate ore with assays of seventy-five to eighty-six percent lead carrying an additional thirty ounces of silver a ton. When he had a shaft down only twenty-five feet, he disposed of a fourth interest in a vein eight hundred feet long and four to forty feet wide for $2,500. His discovery preceded another major lead-silver strike on the Little Lost River side of the towering Lemhi Range in September 1882. Located upon a vein that eventually was traced back to Spring Mountain and Gilmore, this bonanza—like the much larger properties at Gilmore—could be worked only on a modest scale until improved transportation offered an opportunity for extensive mining.

Substantial capital investment, essential to develop all these new lodes, came to Spring Mountain in 1882 as well as to the Viola discovery on Birch Creek not long after. A thirty-ton smelter was installed at Spring Mountain in 1882 at a cost of $135,000. A three-day test run late in 1882 led to improvements so that ore could be processed the next year. But until 1888, miners along that part of the Lemhi Range had to depend primarily upon smelting facilities eventually available at the Viola.

Technology and capital from mines in Colorado provided for improved operations at the Viola. After 5,000 to 7,000 tons of lead ore had been hauled out to Kansas City and Omaha from 1882 to 1885, a smelter was constructed at Nicholia to process the ore locally. About $1.4 million worth of lead and silver came out of Viola in 1886-1887. British capital was introduced in August 1886, with twenty percent dividends paid annually at first. Stock in the London Viola Company more than doubled in value by the end of the first year. After a twenty-two percent dividend the second year, Viola stock values began to decline. Most of the proceeds of the mine (seventy-three percent) now went to pay operating costs. Efforts to cut mine wages failed after a labor strike. A collapse in lead prices in 1888 made matters worse. Then a fire burned the hoist and shaft timbers. These adversities might have been overcome eventually. By 1888 a 1,200-foot vein, worked to a depth of 100 feet, had produced around $2.5 million. At that point, however, the mineral vein was cut off by a fault. Out of ore, the company had to give up late in 1889. About a third of the Viola investment had been recovered in dividends, so stockholders who had hoped to continue to realize a twenty percent return each year wound up losing two-thirds of their capital and gaining no profit whatever. This disaster gave the region bad publicity. The smelter was hauled away, and London investors complained bitterly. Gilmore and Little Lost River mines, which had been shipping a limited amount of ore to the Viola smelter at Nicholia, were also affected by the Viola collapse.

After the Viola shut down, the Gilmore remained dormant until 1902. Little Lost River, which had provided a few hundred tons of high-grade ore to the Viola smelter from 1886 to 1888, did not resume production until 1906. Millions of dollars of Lemhi ore could not be handed, primarily because of transportation difficulties. In 1902, F.G.

Laver of Dubois, Pennsylvania, got interested in the Gilmore area. An investor rather than a miner, he noted a remarkable similarity between the Lemhi lodes and a highly productive mine in which he had an interest in Tintic, Utah. Acquiring a major Gilmore property for a "trifling sum," he joined some associates from Dubois in developing a paying mine. By 1904 his Gilmore mine yielded 2,000 tons of high-grade (forty percent) lead and a fair amount of silver. A slightly higher production followed the next year, but the limitations of wagon transportation for an eighty-five-mile haul to a rail line at Dubois, Idaho, limited the amount of ore that could be processed. Freighting payments for the ore wagons alone ran $10 a ton, so low-grade ore could not be handled at all. Between 1902 and 1908 Gilmore shipped 325,000 ounces of silver and around 6,720 tons of lead bullion in spite of the transportation problem. In 1906 a large steam traction engine with four ore cars (sixty tons capacity) was tried out. On the trip from Dubois to Gilmore, the engine hauled coal, leaving it at refueling stations along the road. Then on the return trip, coal was loaded back onto the steam engine at each of the stops. This imaginative system failed when the four ore cars wore out after a dozen trips. At this point the Pittsburgh mine (Gilmore's largest) suspended shipments and decided to build a railroad. The lack of freight teams cut metal production from the other properties in half the next year. Most of the ore wagons were about worn out, and low prices induced by a national financial panic almost halted production in 1908. A smelter capable of serving a number of smaller mines was erected at Hahn — next to Spring Mountain — in 1909. But that

Only a few (left) of a long line of Viola charcoal kilns (below) survive above Birch Creek

facility ran only seventeen days in 1909 and three weeks in 1910.

Mining at Gilmore finally became practical when the Gilmore and Pittsburgh Railroad solved the freight problem in 1910. Production that year of 5,472,000 pounds of lead and 115,200 ounces of silver rewarded Pittsburgh investors who had induced the Northern Pacific to finance their rail line. In only twelve months of rail freighting, mine production equalled the total attained in all the previous years.

In 1912 the two major Gilmore companies began a joint construction venture for a long tunnel to explore and develop their properties at depth, and the rail line was extended to the portal of their tunnel. This ambitious project resulted in a 6,000-foot tunnel by 1916. The long tunnel attained a depth of 1,000 feet on the lode and made another ten years of mining practical. As a camp of five hundred people, Gilmore had relatively stable production until after 1919. Then a post-war financial and price collapse greatly restricted output until 1924. By that time Gilmore had 20,000 feet of tunnels and shafts and was ready to resume full-scale operation.

Even in mines not favored by rail transportation, high wartime prices allowed profitable operation. Lead-silver mines in the Lemhi Range on Little Lost River turned out around $1 million at a $100,000 profit even though two million pounds of lead had to be hauled over a forty-mile wagon road to a rail connection at Arco. A mill fire in 1918 set this district back, prior to the general collapse of lead mining in those parts a year or two later. But in spite of these obstacles, the Little Lost River lead-silver mines yielded more than $2 million.

Until 1929, when a power plant explosion led to the suspension of large-scale mining, Gilmore remained Idaho's largest lead-silver mining camp outside of the Coeur d'Alene region. Lead mining could not be resumed during the Depression, and Gilmore ended up with a total production of $11,520,852. The Gilmore and Pittsburgh railway continued to serve Salmon for more than another decade, but highway transportation eventually displaced that relic of mining at Gilmore. One of Idaho's better ghost towns remains there as a reminder of the days when Pittsburgh capital developed a major mining camp in the Lemhi Range.

MULDOON

A year after a notable mining rush to Wood River had brought thousands of fortune hunters to search for lead-silver lodes, Jesse Elliott located a Little Wood River prospect which expanded lead production to Muldoon. Following Elliott's discovery in May 1881, promotion in Philadelphia brought eastern capital to the area early in 1882. Muldoon obtained a post office on February 15, 1882, and townsite lots were sold to eager settlers on April 13. By May the incipient city had two tents and a cabin. As soon as a road was completed early in May, freighters began hauling in a smelter. Twelve tents, housing three stores, three restaurants, and three saloons marked the progress of civilization in Muldoon as soon as a road was opened. By late May, Muldoon had a dozen saloons to accommodate a population of almost five hundred.

Forty men immediately went to work on a smelter for the same Philadelphia company which had opened one in Ketchum a year before. Others dug charcoal pits and built a sawmill. Once equipped with a sawmill, Muldoon gained permanent buildings. Much like Ketchum, Muldoon began as a modern, progressive camp. Its Philadelphia smelter followed Ketchum's plant in having electric lights — an installation which preceded those of Hailey and Boise. Mining commenced by mid-June. After four hundred tons of Fish Creek iron ore arrived to enable smelting to proceed, the processing of lead got under way after more than a month's delay. Problems in completing a tram accounted for the late start. Two water-jacket furnaces, each capable of smelting forty tons of ore each day, had been brought in from San Francisco's Pacific Iron Works. One, though, would have been enough. So Philadelphia investors went about purchasing more mines to match the smelter capacity. They already had spent almost $75,000 on smelters and a tram system which brought ore down 1,200 feet from the mine to a two-mile road leading to the smelter that started operations about September 30.

After a month's smelting, twenty tons of bullion was hauled to Blackfoot for rail shipment. By spring they had a 3,000-ton ore reserve. When lime became available to repair enough charcoal kilns to enable smelting to resume, additional production became possible.

In 1882 frontier life in Muldoon had serious drawbacks. On August 27, after unsuccessful complaints concerning the company boarding house, miners discarded their Sunday dinner into Muldoon Creek and insisted that a new cook be employed. The cook was not amused, and company officials responded that no one else could be found to take on that job. On September 17 a violent wind blew a lodging house down, demolishing the facility. The owner moved on to Ketchum. In spite of such unpleasant conditions, sixteen men mined there that winter.

In April 1883, J.W. Ballentine came out from Pittsburgh to organize operations at Muldoon more efficiently. That spring he had twenty charcoal kilns in production. He contracted for a forty mule pack train to bring eight tons of upper Fish Creek iron ore six miles each day to a wagon road. Freighters hauled his ore another six miles to Muldoon. By August, seven hundred tons of iron ore and two hundred tons of galena had been packed in for smelting. Low wages held employment down to a small number of miners, largely because little silver was available and lead prices were low enough that operations were marginal at best. By October, Ballentine's cost ratio turned negative. Running out of operating funds, he had to shut down entirely in October. In November he managed to add ten miners to his staff of four. Encouraged by finding some carbonate ore rich in silver as well as lead, they undertook additional development, which incidentally provided an ore reserve for more than four hundred tons that winter. By May 1884 they had blocked out enough low-grade ore to justify an investment in a fifty-ton concentrator. In the spring Ballentine had returned to Pennsylvania long enough to obtain a lease on June 1 that would enable him to enlarge his plant and operate Muldoon's mine with a partner.

Returning from Philadelphia in June, Ballentine arranged to have his concentrator fabricated in Hailey in order to save time and obtain better service. With a concentrator to supply his second smelter, he anticipated running at full capacity — a welcome change. Modest development work on other neighboring mines promised additional ore to help make his enterprise a success. None of these operations transpired, though. A year later, after his enterprise failed, Ballentine arranged to resume his management position for Muldoon's Philadelphia smelter. By October 1885 he completed a large transportation project, hauling all his tram and mining equipment from Muldoon to a more promising North Star lode on the East Fork of Wood River. In 1886 he went into cattle ranching at Muldoon. The Philadelphia Mining and Smelting Company did not give up altogether, however. The company proceeded on December 6, 1886, to patent several Muldoon claims.

A year later, S.S. Wilson revived interest in Muldoon with his Black Spar mineral discovery on Bear Gulch. Selling part of his interest for $30,000 to St. Louis investors, he managed to employ ten miners on January 23, 1888. Experience as a California placer miner from 1849 to 1855, and as a quartz miner since then, helped him promote his new property. By May 1, 1888, he had three hundred tons of ore ready to mill and two more carloads ready to ship to a smelter. His ten miners kept on working through 1888, but declining lead-silver prices discouraged an effort to construct a concentrator. Operations were shut down altogether from December 1888 until April 1889. Although a sawmill was brought in to Muldoon, 1889 proved to be a pretty quiet year until a vast forest fire in August burned out the smelters (which had not operated since 1884 anyway) and houses. Some development work continued on smaller Muldoon properties that fall. But irregular ore shoots in complex fissure veins made development very difficult. The inability to operate a smelter without a greater variety of ore — which hardly could be imported into an isolated camp like Muldoon — compounded the problem of trying to mine there. Since Muldoon's lead-silver lode extended for several miles, a number of companies continued to devise a satisfactory technology for operating. They failed repeatedly.

⚒   🐎   ⚒

Undeterred by disaster, promoters of Muldoon properties kept up an attempt at mining. J.E. Smylie's Lake Creek bonanza discovery east of Copper Basin on upper Lost River, only nine miles from Muldoon, encouraged a mining revival in September 1891. By the spring of 1892, several Muldoon properties had accumulated modest ore reserves. Yet the miners needed a direct road to a smelter in order to operate. In spite of the problem, enough Muldoon lodes looked good enough in 1892 that William Hyndman took over the Philadelphia mining property the next year. Frank E. Johnesse tried to promote another group of new Muldoon claims in 1896. By that time, Muldoon was being viewed retrospectively as a famous old producer. Muldoon's greatness, unfortunately, derived from magnificent production that had been anticipated rather than achieved.

Development activity in 1901 to 1902 brought renewed promise to Muldoon. Finally in 1906, about sixty miners revived that unfortunate camp. Then a spectacular catastrophe created some genuine excitement:

> During the night while no one was working, a great body of water broke into the tunnel, presumably at the face which had just passed a porphyry dike, and the flood continued for about forty hours under enormous pressure. The pressure and volume was such that it washed away the car, tools, shop, and a big dump which

had been accumulating for a year, as if by magic. Just below the dump a grove of large fir trees, some of them three feet in diameter, were cut out and uprooted by the escaping flood and tossed aside like straws. It cut a gulley down the side of the mountain to bedrock in places thirty feet deep, rolling over and pushing aside boulders of many tons weight in its course. It raised the creek out of its banks for four miles. The water was discolored and muddy like yellowish tailings. After forty hours the flow gradually decreased until at the present time there is a stream about one and a half inches deep flowing out of the mouth of the tunnel. The tunnel is filled with mud and debris that tapers back to the roof at a point about seventy feet in beyond the entrance. Many pieces of what appears to be calcite casing and quartz crystals are mixed with the debris in the tunnel together with pieces of galena and carbonate ore. A spring which was formerly flowing on the surface nearly over this tunnel and which had been drained by it, has commenced to flow again, which leads to the assumption that the underground reservoir is not yet completely drained but simply choked and dammed up and when open will flow again. A crew of men have been put to work to clean up this tunnel and underground mystery.

Work at Muldoon went right on after this mishap. The same month Robert T. Tustin, who kept Muldoon active until 1912, started building cabins for an extensive new operation. Supported by Arizona capital, he started bringing in a new one hundred-ton mill the next spring. The lack of a road better than an existing steep grade from Bellevue held him back: a four-horse team could not haul more than 1,500 pounds up the hill. On May 26, 1908, he got support in Hailey for a superior alternate route. Construction began in June, but in August this ambitious Hailey project had to be suspended for lack of funds. Five hundred dollars was raised to finish a one-lane road with turnouts, but $200 more was needed. Eventually Hailey's road was completed in November, just in time to be closed by snow.

Tustin finally went ahead with his mill project without waiting for a better road. In September 1908, he had fifty or sixty miners and builders employed with a payroll of $7,000 to $8,000 a month. Wages ran $4 to $7 a day. With hydroelectric power, his new mill, tramway, and assay office would be thoroughly modern. New bunkhouses were also necessary to accommodate Tustin's large crew. Another property six miles from Muldoon was also active in September 1908. A force of eight or ten men built a new camp there that winter.

Although work had to be suspended for a time in January 1909 because of winter weather, construction was completed in time for milling to commence that summer. During an initial season, Tustin's hydroelectric mill (driven by a 200-foot head of water that ran 2,700 feet through a 22-inch pipe to a generator) was "producing a very high grade product." By September, Muldoon had stage service to Bellevue (since Hailey's road never was reopened), and Muldoon ranchers and farmers profited considerably by having a local market for their products as well as better access to Wood River. Shutdowns in 1910 and 1911 resulted from problems of operating an unreliable, low-grade mine with a cost-ratio rarely more than borderline at best. Only twenty miners even tried to resume work in 1911. If they had not had to invest $8 a ton hauling out their concentrates, they might have had a chance. But late that season, Muldoon's mining machinery and equipment was consigned to a mill between Hailey and Ketchum. In 1912, Tustin decided he would do better trying to revive Bay Horse instead of Muldoon.

With production of only about $200,000 in an initial era (1882-1884) and not much added from 1908 to 1910, Muldoon never lived up to its earlier reputation which had depended mainly upon extensive capital investment rather than the recovery of lead or silver. Attractive enough to be built up twice, Muldoon's lodes deceived many investors whose major contribution turned out to be machinery which became available for other mines near Hailey. Muldoon finally emerged as a superlative livestock country after sheep and cattle ranching, which had commenced there to supply mining markets, supplanted lead and silver as a major element in Little Wood River's local economy.

After mining was suspended at Muldoon, sheep moved in to graze by old charcoal kilns

LITTLE LOST RIVER

A mining rush from Leesburg on June 15-16, 1867, created interest in Little Lost River more than a decade before T.C. Blackburn made a major lead-silver discovery there in September 1882. By the spring of 1883 Blackburn had organized a mining district, recognizing a lode that could be traced for six miles but which outcropped in only a few places. An experienced prospector from Deadwood, South Dakota, he had followed up an interesting Viola lode find that his brother, Charles F. Blackburn, had located across Diamond Peak on adjacent Birch Creek. An initial twenty-five-foot shaft provided access to a drift on a twelve-foot vein that furnished assays ranging from 6 to 1,580 ounces a ton. Early production came from ore hauled to the Viola smelter on Birch Creek at Hahn from 1886 to 1889. With closure of that smelter in the fall of 1889, Blackburn's mine had to suspend operations.

By 1906, work resumed on another property along Blackburn's lode. Only a small crew was employed for a time, although a one hundred-ton concentrator stepped up production for a decade after 1908. After the loss of a hoist, concentrator, and other equipment by fire in 1918, production ceased while a new plant was built. Work resumed in 1922, and a new mill produced continuously from 1924 to 1931. More than $2 million worth of lead-silver finally came from Blackburn's district.

Part IV. Late Nineteenth Century Mineral Discoveries

INTRODUCTION

After 1884, a half dozen promising new Idaho gold camps attracted prospectors or investors from a wide area. Two of them came into production without unreasonable travail. But for one reason or another, difficult obstacles interfered with major production anticipated from four later nineteenth century discoveries. Remote locations proved a severe drawback. Lack of ore — a deficiency crucial for a mining camp — retarded several of these camps. In one area, the availability of only a limited amount of gold delayed large-scale development until another mineral, cobalt, made Blackbird into a major Idaho metal producer. Idaho's last major gold rush brought thousands of miners to Thunder Mountain, but almost all of that district's wealth came from Pittsburgh investment rather than from mineral resources available for exploitation along Monumental Creek. This slightly unfortunate approach to mineral development characterized the mining camp of Graham even more. By 1884, most of Idaho's traditional gold and silver lodes had been located, so most gold rushes after that time were somewhat misdirected. Other metals of great value remained to be found. But truly important undiscovered gold fields, however, no longer turned up to justify old-fashioned mining excitements.

BIG CREEK

Mineral discoveries near Elk Summit high on a ridge between Big Creek and the South Fork of the Salmon River came a decade before prospecting on Monumental Creek expanded Big Creek mining possibilities into an even more remote area around Thunder Mountain. Deep canyons and rough country delayed the development of mining anywhere on Big Creek. A gold rush early in the twentieth century finally brought a horde of prospectors into Idaho's Salmon River mountain wilderness west of Leesburg and north of Stanley and Deadwood.

Antimony had been noticed in the area years before anyone succeeded in identifying commercial gold and silver. A Thunder Mountain lode, which no one could develop, and some Chamberlain Basin placers had been investigated as early as 1866 or 1867. Nothing came from that exploration. Finally James Reardon and L.M. Johnson brought a small discovery party to Big Creek as early as they could prospect in 1884. In June, they found an 1,100-foot outcrop of a system of parallel veins about 60 feet wide. A year later, on June 15, 1885, they organized Alton mining district, and that summer 150 miners located about one hundred claims. They found silver ore described by Norman B. Willey as "refractory, but not base." In 1886, prospect cuts had reached a depth of fifty feet. A.L. Simondi, a Weiser assayer, created a lot of interest when he reported a 2,000-ounce silver sample in August. A ton of ore from these exploratory holes, packed out to a railroad at a cost of $80, provided a favorable test yield of 267 ounces of silver later in 1886. Since an eighty-five-mile wagon road would have to be constructed at an estimated expense of $20,000 to reach the district, miners at Alton faced a severe obstacle. The ore, distributed in small stringers through a broad zone or lode, could yield flattering assays for selected samples, but averaged only a $1 or $2 a ton. A large low-grade lode of that kind eventually could be worked profitably in the twentieth century when newer methods and good transportation were available. Elk Summit offered no such attraction.

The gradual expansion of mining possibilities around Alton, both in the immediate vicinity as well as around Big Creek, came during two decades or more of prospecting. Following the pioneer work of John Osborn in 1880, a modest excitement attracted interest on Sugar Creek in 1887. James Hand located a Beaver Creek claim on August 18, 1893, which he retained for half a century. A more promising find brought more miners to Smith and Government Creeks near Alton in 1898. A Topeka firm acquired this property in 1902 and eventually drove about 2,000 feet of development tunnels in a lode 200 feet wide. Returning to Beaver Creek in the spring of 1899, James Hand

> discovered and located the most extraordinary ledge on the North American continent. It is an enormous porphyry dyke of free milling quartz that stands out boldly like a huge cathedral. Measurements taken show the ledge to be 300 feet at the widest and 60 feet at the narrowest part. The ledge can be easily traced for over three miles.
>
> Assays of the croppings of this ledge made by Mr. Tillson, of the Iola mine, show values ranging from $18.50 to $186.60.

Another nearby discovery of Charles Crown brought miners to Logan and Fall creeks in 1899. Crown went on to find "some remarkably rich locations in Thunder Mountain" that season. But his Logan and Fall Creek prospects proved disappointing. By 1902, about 2,000 feet of development tunnels demonstrated an absence of ore (as evaluated in such a remote area), but after some additional effort at development, George Lauffer and Joe Davis relocated this abandoned property in 1908. Not much aside from negative information came from all that effort.

North of Big Creek, Richard Hunter reported an unexpectedly successful 1899 placer operation:

> In the Chamberlain basin, strikes showing phenomenal values have been made by the Briggs brothers, of Ohio, and a quartet of lucky prospectors from Utah. The Ohio boys located a placer claim on the top of a mountain and worked like Trojans for two weeks to the intense glee of the old rock smashers. The boys succeeded in getting a 12 hour run of water and washed out $1,876 in coarse gold. In the clean-up nuggets

Big Creek (c. 1900)

worth $10 were found. The hilarity of the 'waybacks' ended suddenly.

Copper also created excitement in 1899:

> Mike Nevins, the genial, big hearted proprietor of Nevin's cosy ranch, at the mouth of Elk creek, has located a colossal ledge of copper near the fork of Elk and Smith creeks. As the ledge towers upward to a height of over 600 feet the reader can form a slight idea of the magnitude of Nevin's discovery. A representative of Marcus Daly has gone to examine Nevin's discovery.

A somewhat more successful effort attended another nearby discovery of 1903. Four years later a small three hundred-pound prospect mill turned out $173 in a seventeen-day run. A five-stamp mill, brought there in 1911, produced a $6,000 or $7,000 yield by 1916. In addition, a fourth Alton lode discovery on Government and Logan creeks filled in some mining territory between the 1898 and 1899 segments. In 1911, D.C. MacRae and E.F. Goldman located claims along a ridge between Government and Logan creeks, but they had low-grade ore at best. Some may have gone as high as $4 a ton higher up in their vein and $2 at greater depth, but their average ran lower. Development of this series of four mining areas along a single northeast and southwest mineral zone showed that a large lode extended close to four miles in length and one hundred to three hundred feet in width. Yet almost no production could be managed at such a difficult location. During the Thunder Mountain rush, some of these properties

> acquired an unenviable reputation by reason of unwarranted wildcatting operations of that period, but not a single instance of intelligent mining development was then recorded, and as a matter of fact 90% of the money raised from the sale of stock based on Big Creek properties during that period was used for promotion purposes and never reached Idaho.

Farther down Big Creek, other lodes had more of a chance for development. W.A. Edwards' property, located in 1904 on a ridge between Logan and Government creeks (below D.C. MacRae's later discovery in 1911), justified importation of a stamp mill. Logan City (later Edwardsburg) began that summer with a saloon, store, butcher shop, and a house on Big Creek flat, and a four-stamp mill arrived in 1906. Milling finally began five years later, with a larger mill imported in 1909-1910 that produced $1,200 in 1911. Sulfide ores, requiring a cyanide process, continued to present a problem which accounted for so long a delay and such small production. Edwards also held additional claims twelve miles farther down Big Creek, where a 2,500-foot lode was developed. A series of Ramey Ridge discoveries in 1908 led to mining expansion there. Most of Big Creek's production came from the Snowshoe mine in that area, with a yield of about $400,000 between 1906 and 1942.

GRAHAM

Prospectors in search of lost north Boise mines had known of large but otherwise unpromising veins in Silver Mountain for twenty years or so before Matthew Graham's careful examination of some dull red outcrops created interest in the district late in 1885. Graham, a well-known Atlanta miner, felt that, although the ledges of Silver Mountain had attracted no attention in the past, they offered great possibilities in the future for large-scale quartz mining. Assays of some good samples confirmed his expectation. Only the deep December snow held back a rush to Silver Mountain.

Matt Graham had gained long experience in promoting important Atlanta mines in New York and London, and for years had spent most of his winters in New York City, where his close resemblance to Congressman William Marcy Tweed still was noticed long after Tweed died in prison, where he had been sentenced following the exposure of the notorious Tweed ring. Now Graham set out to develop Silver Mountain. His friends in Atlanta (only sixteen miles to the southeast) responded immediately. An Atlanta newspaper correspondent foresaw on December 30 that

> the North Boise mines will draw thousands of miners, prospectors and capitalists. The tin horn gamblers will be there to work the greenies. There are ledges out there looking far better than the Custer mines did in 1878, since when $2,500,000 have been taken out.

When Graham reached Boise and reported that he had veins four to six feet wide with surface assays

of $50 to $2,000 of free-milling gold and largely metallic silver for which no complicated reduction process would be required, the *Idaho Statesman* concluded on January 2, 1886: "It is evident that the new discovery will eclipse any of the older quartz discoveries in Idaho." And the *Atlanta News* (quoted in the *Boise City Republican* on January 23, 1886) spoke even more enthusiastically:

> The ledges in Silver district are simply enormous; they vary in width from ten to three hundred feet, cropping out like gigantic walls to protect their wealth, and can be traced for miles.

Exploration of the most promising lode on Silver Mountain commenced early in 1886. Two shifts of men drove a tunnel 240 feet to strike the vein about 200 feet below the surface, with results favorable enough that Matt Graham managed to interest London capitalists in supporting his new Idaho Gold and Silver Mining Company, Ltd. By the fall of 1887, a $15,000 road was completed to the camp and a 500-foot exploratory tunnel (along with a 112-foot inclined shaft) was run into the lode. The lode, 30 to 40 feet wide where the vein struck it, contained what was interpreted to be ore worth $30 to $50 a ton, with a richer zone 6 feet wide running at $90. Encouraged by such a development report, the company began to build an elegant twenty-stamp mill. About 150 men were employed. (Wages of $4 a day for miners, $3.50 for outside workers, and $7 for carpenters and stone masons were unusually high for the time; these rates reflected the difficulty of getting skilled labor to work in the remote district.) Fifty or sixty men worked right through the extremely hard winter of 1888, and reports of rich new strikes made for great enthusiasm in the new camp.

In the summer of 1888, the Silver Mountain boom reached its height. The new town of Graham boasted of having

> six saloons, one store, five boarding houses, one restaurant, two blacksmith shops, a jail, a Justice of the Peace and Deputy Sheriff, one butcher shop, two faro games, three livery stables, a fine hall, 300 men, forty-one ladies, and the controlling vote of Boise county.

George M. Parsons, the mine superintendent, decided that the jail—"a strong affair of logs, nails, planks and iron"—would have to be erected after numerous incidents of assault and battery and larceny insured that there would be "quite a number of guests" sent there by Justice James D. Agnew.

By August 12, the mill was completed. A mile-long tramway to haul ore from the mine to the mill went into service. Telephones connecting the mine and mill were installed, and in spite of its wilderness location, Graham was well supplied with the conveniences of civilization.

All that the promising new mining camp of Graham lacked in the late fall of 1888 was ore that could be treated in the excellent modern mill which had been installed at an expense of $350,000 in that difficult location. After a few trial runs, the mill shut down. Matt Graham spent the winter in London arranging for more British capital to develop the necessary orebodies. Mining at Graham had been commenced on the theory that, although surface prospecting was generally disappointing (in spite of some rich samples that could be found in the outcrop), development at depth would reveal enormous, rich orebodies. When the lode could not be worked at the depth tried originally, Graham wanted to drive a 5,000- to 6,000-foot tunnel at much greater depth in the summer of 1889. But while Graham was on his way to London, an attachment for unpaid debts led the county sheriff to take over the administration of the mine on November 23, 1888. Those who spent the winter in Graham had plenty of leisure: only the watchman had anything to do, and the watchman's job really was not very hard.

Three English mining engineers completed an examination of the property on June 10, 1889. Because they were "very much pleased with the mines," they recommended that Graham's 6,000-foot tunnel be driven. But London capital, already invested to the amount of $600,000 in the north Boise mines, was not forthcoming.

In three sheriff's auctions—one at Graham on August 31, another at Rocky Bar on September 10, and the other at Idaho City on November 16—the Idaho Gold and Silver Mining Company's property was sold to satisfy the unpaid claims of creditors. In the final sale, the $350,000 mill went for $9,500, and the tramway, buildings, and the thirteen mines (thought to be the most valuable silver properties in Idaho only the year before) realized only $500. Altogether it was estimated that about $1 million was expended to prove that gold and silver ore (mineral deposits which can be worked profitably) was lacking entirely at Graham. Yet that prospect retained genuine possibilities for development, and after 1980, efforts to discover a large ore body were resumed.

PINE GROVE, BENNETT MOUNTAIN, LIME CREEK, AND WOOD CREEK

Extensive prospecting along the South Fork of the Boise River had gone on for more than two decades prior to the location of the Franklin lode at Pine Grove on May 16, 1887. Oliver Sloan, who had lived for years at Pine Grove, got a number of prospectors to explore lode possibilities there in the summer of 1886. The delay in finding anything came from lack of ore in their outcrop. Finally the prospectors struck a good vein in a blind tunnel. From then on, exploration came quickly.

By the spring of 1888, lodes at Pine Grove attracted a considerable gold rush. A St. Louis company built a stamp mill that summer, and others were brought in from Hailey and Rocky Bar. Sloan managed to sell part of his interest for $45,000 and to develop a town of two hundred to three hundred miners. Four saloons, two stores, a restaurant, and two barber shops served his new community that spring. High-grade samples from the Franklin assayed from $300 to $900 a ton. More important, a fourteen-day mill run for the St. Louis company, averaging fifteen tons a day, returned a $20 to $30 profit per ton. A neighboring mine tested at almost $90 a ton at a Rocky Bar mill. Forty men were employed in developing the Franklin that summer. With British investment in 1892, Pine Grove took advantage of national economic depression that favored gold mining for several years after that.

Although not an enormous producer, a ten-stamp Franklin mill eclipsed all other Elmore County mining operations (including Rocky Bar, Atlanta, and a number of other later discoveries) for several years after a $20,000 sale in 1897. Running night and day when not broken down, the modest operation could produce $100,000 a year. Even "with its little old 10-stamp mill hung up half the time for repairs," the Franklin managed to recover $70,000 in 1904. The addition of a one hundred-ton cyanide plant to process a 2½-year accumulation of tailings (worth $7 a ton) helped out still more in 1905. The loss of the entire $20,000 plant in an unfortunate fire on May 16, 1908, did not help, but miners at Pine Grove finally produced about $750,000 in gold.

Additional mineral discoveries farther down the South Fork attracted investment to some Bennett Mountain properties near Dixie in 1892. A Salt Lake City purchase of a $50,000 half-interest brought in essential development capital in 1894, enlarging minor development operations financed by a small quartz mill's yield. A large 60-foot vein of $20 ore showed great promise after a 210-foot development shaft was sunk the next year. By 1896, 75,000 tons of $14 ore had been blocked out. Within a decade, a 300-foot shaft had yielded several car loads of high-grade ore. In 1904 a 4,200-foot tunnel was commenced in order to reach a depth of 1,100 feet on the vein. This project proved overly ambitious, however.

Between Bennett Mountain and Pine Grove, a

number of additional lodes gained prominence along Lime Creek. A 200-foot tunnel exposed three low-grade veins at the Hawthorne, where a ten-stamp mill was under construction in 1903. Two other Lime Creek properties were developed by small companies with Salt Lake City capital, with a twenty-five-ton roller mill brought in for one of them in 1904. Then an 1,800-foot Copper King out-crop was traced on Wood Creek the next season. With additional finds, an extensive area between Pine Grove and Dixie had quite a number of interesting possibilities, which did not become major producers but which kept quite a number of prospectors and developers busy for many years.

NEAL

Cattlemen and packers had spent years on upper Blacks Creek near a summit that provided access to Willow Creek before any of them noticed outcrops of important gold lodes. Jack Slater, in fact, had cut right through a mineralized outcrop while building a road, but "tossed it aside as worthless rock." Tom Johnston of Caldwell picked up some ore in 1887 but lost his samples before he could get them tested. Then in December 1888, Arthur Neal came across some rich float while bringing his pack string from Pine Grove out to Boise by way of Willow Creek. Since he could not prospect that late in the season, he spent the winter in Mountain Home and returned in the spring to try again. On July 20 he found what he was looking for while searching for water for his pack string. His mineral discovery outcropped too high in elevation to have much water handy. That fall, in partnership with George

Thomas Johnston (right) of lower Boise discovered Neal mining district, bus mislaid his ore samples before getting them tested.

House, he began to develop a valuable lode. A number of other prospectors found additional mineral evidence nearby. In September, T.H. Callaway (a prominent Confederate refugee from Missouri who had spent more than two decades in Idaho) recommended Neal's mines as the best he had seen. Placer deposits, also available there, could not be worked for lack of water. But lode mining at Neal had a really bright future. One owner refused a $2,500 offer for an undeveloped prospect, and others showed equal confidence.

In common with most Idaho lode districts, Neal's mines attracted eastern capital. Four major properties owed their development to outside investors, mainly from cities like New York and Chicago, who brought in mills and engaged in essential exploratory work. In at least one instance, a local owner managed to develop a valuable property without access to major outside capital resources. Yet after that unusual achievement, he sold out to a Chicago and Wisconsin investment group in 1902 for $225,000. Within a year, 15,000 tons of ore (seventy-five percent free-milling and twenty-five percent susceptible to cyanide recovery) worth $16.20 a ton had been blocked out, so no loss would be incurred in that transaction. Another property produced $40,000 in 1904 in operations incidental to development, and a third mine had a two hundred-foot shaft and several thousand feet of tunnel to expose $16 to $18 ore in a vein up to sixteen feet wide. A long development tunnel opened a fourth property, and new mills were being brought in to increase production. Altogether, mining at Neal worked out satisfactorily: before operations were shut down, more than $2 million came from the district.

BLACKBIRD

Discovered in the summer of 1892 by Indian Tom, a local Lemhi prospector who brought rich float copper samples of his find to Robert N. Bell in Salmon, Blackbird began to attract attention several years later. Bell and Indian Tom set out on October 17, 1893, to examine Blackbird's lode possibilities, which Bell confirmed as a rare instance in which an Indian mineral report could be veri-fied. During further work in the summer of 1895, a free-milling gold lode at Blackbird soon was traced for 2 to 3½ miles along the surface. J.O. Swift found this lode attractive enough to invest $12,000 in a half interest and to promote his new property in the fall. Development work undertaken by twenty miners in the winter showed a change in the ore at a depth of fifty feet all along the lode. Gold values

The substantial cabin of a miner survived many winters near Cobalt

ranged from about $12 a ton on the surface to $24 farther down. (These rates compared favorably with nearby Yellow Jacket.) Then at fifty feet down, their lode changed to copper which retained gold values as well. By 1896, Blackbird had been established as a promising mine with ten to thirty percent copper that had to be smelted. A Rothschild investment firm in London agreed late in 1896 to undertake development and to purchase J.O. Swift's Blackbird property for $250,000 if their own exploration justified such an investment. After a year's investigation, gold, copper, nickel, and cobalt valued at $2 million was identified. The isolation of Blackbird from rail transportation, the lack of a sampling works, and the difficulty in processing ore all contributed to long delays in serious production.

After additional claims were consolidated in 1899, a modest production of $35,000 in copper was finally realized from 1913 to 1915 and in 1921. Then cobalt became more prominent after modern production got under way in 1939. Active until 1960, Blackbird accounted for fourteen million pounds of cobalt of which most was produced from 1952 to 1959 under a government contract price of $2.30 a pound. In a ten-year span, 1949 to 1959, Blackbird became a major Idaho mining area, credited with $47.5 million in cobalt, copper, gold, and minor amounts of nickel. Rising prices allowed Blackbird to resume production in 1967 with $1,186,000 primarily in copper turned out that

year.

The unreliability of African sources of cobalt and dramatically increased world prices associated with the instability in foreign mining areas led to a resumption of interest in Blackbird in 1978. Activity resumed there in 1980 in order to achieve a 2,000-ton-a-year production level by 1984. With development costs projected at $230 million — and with $35 million already invested in mine rehabilitation, water treatment, and mill testing — operations were slowed down in November 1981 pending a government contract to guarantee an adequate price to cover production expense. A staff reduction from 105 to 80 came at that time, but a projected increase to 650 employees, including a Blackfoot smelter operation, was contemplated in 1982. But failure to gain government price support led to an additional staff reduction early in 1982 with 10 of 63 miners dismissed then. Concentrates still were milled at a rate of three tons an hour for future testing after a refinery was constructed, but even that level of activity was curtailed somewhat pending federal guarantees.

THUNDER MOUNTAIN

Like most other western mining states, Idaho definitely needed the excitement of one or two big gold rushes at the end of the nineteenth century. People were restless. Frontier opportunity to start a small farm was limited. Although most of Idaho's farm land still awaited development, large irrigation projects, as yet only contemplated, had to be undertaken. Those who preferred mining had fewer options than were available forty years before. Occasionally some one tried to set off an Indian war, but most Indians no longer had much interest in that kind of diversion. Some Idaho miners got mixed up in the antecedents of the Boer War in South Africa but that was all remote. When the Spanish-American War came along in 1898, as a partial satisfaction for this need for excitement, only a limited number of Idaho volunteers had a chance to participate. As for gold rushes, Cripple Creek offered a good model in Colorado. Many Idaho miners also joined the Yukon gold rush to Dawson. But those places were far away, and the excitement there did not infect Idaho. Idaho did not even have a state department of gold rushes, although a state mine inspector's office met much of that need. Whenever a halfway suitable gold discovery should happen to come along, enough potential miners were ready to furnish more than the appropriate amount of excitement. As matters

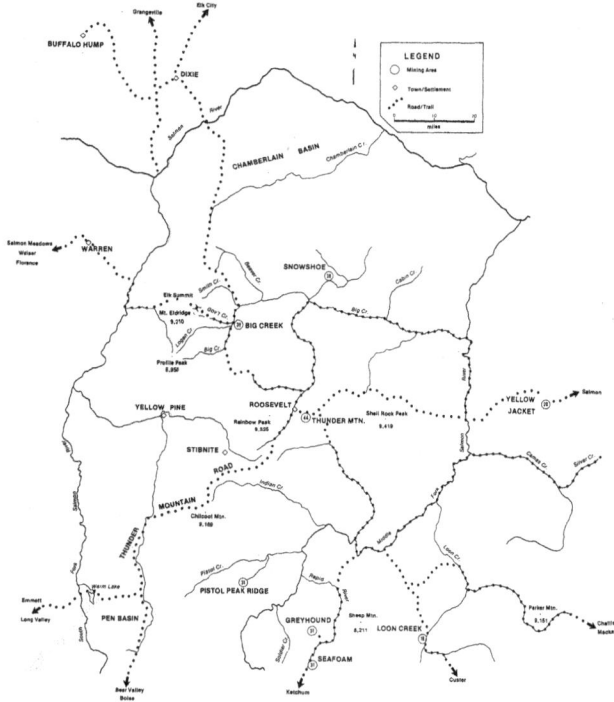

Thunder Mountain and nearby mines. Numbers in circles refer to index map of mining areas, page 2.

turned out, Buffalo Hump, north of the Salmon River, and Thunder Mountain, farther south, offered a much needed opportunity for anyone predisposed to join an old-fashioned gold rush. The two areas provided a fitting climax for Idaho's nineteenth century mining history.

Locating two lode claims on Buffalo Hump on August 8, 1898, miners from Colorado set off a wild gold rush in the fall and winter to that forbidding ridge after almost four decades of prospecting had disclosed nothing of great interest. Thunder Mountain discoveries had already been reported, but Buffalo Hump diverted attention from rival prospects for two or three seasons. Finally Thunder Mountain was the object of a rush comparable to the one at Buffalo Hump. Neither district actually produced very much gold. Still, they made up for the lack of mineral wealth by entertaining fortune seekers and investors on a lavish scale. They contributed far more than their share to developing Idaho's economy, even though they failed to advance mining to any really appreciable degree.

As Idaho's final big gold rush, Thunder Mountain had antecedents which went back to another era. Earlier prospectors radiating out from Warren after 1862 were attracted by Thunder Mountain's conspicuous mineralization, and wild tales of those early investigations had wide circulation after 1899. Other accounts had greater accuracy. James W.

Poe, a prominent Lewiston pioneer of 1861, discovered a good outcrop with free-milling gold on Thunder Mountain in 1866 or 1867. That led him to search "for a placer ground that would apparently go with the rich quartz lead. But the country was then the summer stamping ground of the Sheepeater Indians, who about this time became troublesome, and white prospectors were compelled to leave." Poe reported that Chamberlain Basin miners also had to evacuate because they could not process gravel worth less than 25 cents a pan in 1867. On a trip to Buffalo Hump in 1899, Poe returned to Thunder Mountain to find that his old discovery had been taken up during an early phase of mining development there. Poe was fortunate not to get too involved in Thunder Mountain lode properties, although some of his unpromising 1899 claims were later relocated and sold to Pittsburgh investors.

⚒ 🐴 ⚒

Although Poe could make no use of an almost inaccessible gold lode at Thunder Mountain in 1867, mining conditions had improved in three decades. Ben and Lou Caswell, twin brothers from Michigan who had learned something about prospecting in Colorado, had searched for Seven Devils mineral wealth with no success at all in 1894. During the Panic of 1893, gold mining was favored over copper, and they decided to hunt for gold in as remote a wilderness as they could find. By then they were broke. The way Ben reported it, all they "had was a bunch of scrawney cayuses — in fact they represented about our only possessions when we went into the Seven Devils, so we can't say we lost anything there." Finding good surface indications on Thunder Mountain in August 1894, they settled on Cabin Creek, trapped and hunted for a living, and came back the next two summers to use rockers during short two-week seasons when placering was practical. They recovered $245 worth of gold in eight days in 1895 and another $190 in 1896. In 1896 they spent most of their time whipsawing lumber for sluices, so as to increase future production. Then their brother Dan and his partner, Wesley Ritchie, came over from Montana to join them in producing $900 in 1897 with sluices. Encouraged by this success, Ben and Lou came out to Boise on August 10 with a remarkable story:

> That there are Klondikes yet hidden from the knowledge of men in the wilds of the Idaho mountains was demonstrated yesterday when the discoveries of A.B. and L.G. Caswell in the Salmon river country became known.
>
> These men came into town with a large clean-up of gold. When asked about their dis-

covery they stated they mined the gold on what they called Mule creek, which heads in a mountain which they have named Thunder mountain. Mule creek flows into Monumental creek, this into Big creek and Big creek into the Middle fork of the Salmon.

The brothers discovered the claims some four years ago. The first three seasons they made expenses and this year they have secured a fine clean-up. They expressed the belief that the district will make a good camp; and from their report of the character of the discovery that belief would seem to be well founded.

They have been placer mining the surface of lode claims, working the debris on the mountain side. The entire mountain, they say, is gold-bearing and the gold they have been getting has been released by the decomposition of the formation over which it is found.

This remarkable mountain is porphyry. The prospectors describe it as being a great volcanic crater which has been filled with the gold-bearing rock. The gold is found everywhere on the mountain. The brothers have prospected the ground very thoroughly and say they have pounded up fully a ton of the porphyry. It all pans well. In addition they have had a number of assay tests. The result of their investigations is such as to convince them that the mountain is an enormously valuable deposit of gold. They do not claim the rock is high grade, but they are well satisfied with its value. They did not care to state what the average value of the rock is as determined by their investigations.

Their mountain of gold, even though low-grade in value, contained more high-grade pay streaks than they had noticed at first. Yet their massive-appearing gold deposit had geological characteristics not typical of ordinary mines. They were dealing with an unstable formation which confused more than a handful of competent mining engineers.

Slides of soft, disintegrated rhyolite which absorbs water in a wet year and begins to roll silently along slick faults, have occurred frequently on Thunder Mountain. (One in 1909 which plugged Monumental Creek and flooded the town of Roosevelt was the most recent of many in the area.) These slides are of some commercial importance, since they contain the gold. Prospectors and miners dug around in the mud flows, since gold was precipitated rather widely over the surface in a manner most extraordinary for gold mining. The remineralized mud flows had some especially rich gold deposits on the surface—gold, in fact, had even precipitated on old (thousands of years old, that is) wood chips in really a most irregular manner. Prospectors got the notion that Thunder Mountain (or at least the rhyolite mud slides on it)

was a mountain of gold; there was enough solidification of mud slides from the action of salicic acid as to give these essentially placer deposits the appearance of soft rock which might be worked by quartz methods. The gold (already precipitated by carbon through a process much like that of a ball mill) was easily recoverable, and the enriched surface concentration of the hillside mud placers gave an entirely false impression of the extent and richness of the district.

In the early phase of Thunder Mountain's mineral development, some less scrupulous promoters contributed even more in the way of artificially brilliant gold assays. Salting the samples ("salting" is the deceitful process of enriching a sample after it is dug by slipping some already mined gold into the test before assaying) was more than ordinarily common: John Oberbillig, for example, assayed there in 1904 and caught some of his clients salting their samples so thoroughly that even barren bull quartz would go $20 a ton. Yet natural surface enrichment in the initial claims gave Ben and Lou Caswell good enough returns that they had no need to resort to fraud. They were not trying to peddle their claims, since they were gaining handsome returns for a very short mining season each spring when melting snow provided water for a few days high on the mountain. They had no way of anticipating that their mountain of gold was actually a mountain with a gold skin. Before this unusual deposition of gold could be evaluated accurately, however, they got involved in one of Idaho's wildest mining developments.

⚒   🐴   ⚒

Returning to their ranch on Cabin Creek, Ben and Lou Caswell prepared for an enlarged operation in 1898. Reinforced by another brother, Dan Caswell, and his Montana partner, they obtained about $3,000 in their two-week season and managed $4,000 during a similar run with a small hydraulic giant in 1899. Because they could sluice only a short time while melting snow provided water on the mountain, they managed only brief annual seasons. They received publicity sufficient to attract a number of prospectors in 1899. That fall, S.W. Emerson reported in Grangeville:

> The ledges are composed of talc throughout which are found kidneys of sulphide ore that goes $3600 tr [sic] the ton. The talc is dug out and allowed to slack one year and is then run through sluices which virtually make a placer mine of a quartz proposition. The tailings after slacking from ten to twelve months are put through the sluices, giving good return for the labor

expended. In fact the tailings sluiced the third time will pay expenses. A stamp mill would save nearly all the values, it being strictly a free milling ore.

H.E. Taylor, impressed by the geological similarity between Thunder Mountain and Colorado's Cripple Creek district, where he was an experienced mining engineer, concurred with Emerson. After a Caswell recovery of thirty pounds of gold in only forty-two hours of operation with a small hydraulic giant in 1899, he anticipated that with "immense richness" and "phenomenal surface showings," Thunder Mountain was "destined to be the leading gold camp in Idaho." Efforts to interest substantial investors proved more difficult, however. Taylor organized the Thunder Mountain Consolidated Gold Mines Company with Weiser capital. His plans to bring in a twenty-five-ton Huntington mill in 1900 proved to exceed his investors' resources, and his venture collapsed.

A search for wealthy mine owners eventually produced better results. J.R. DeLamar, who had gained a fortune in developing large-scale mining near Silver City, had Thunder Mountain checked out by engineers prudent enough to keep him from going into a losing proposition. W.H. Dewey, another large Silver City operator, was less fortunate. His son, E.H. Dewey, had excessive confidence in Thunder Mountain. In 1900, after seeing a $500 short run Caswell production, W.H. Dewey agreed to purchase the property for $100,000 if a full investigation should warrant such an investment. Dan Caswell certainly had a good operational report:

> In a run of 72 hours this season, with hydraulic power, they took out 29 pounds of gold, avoirdupois weight. The gold is worth $13 an ounce. The 29 pounds was worth between $5000 and $8000. They worked this year an area of 75 feet square and to a depth of three feet. These placers are nothing more nor less than a huge porphyry belt. The material is dug out and slacks, after which it is washed. Below a depth of three feet it is too hard to work by this method, but carries as much gold as from the surface to that depth. The extent of this auriferous porphyry is not known, but the whole mountain appears to be porphyry. Three tons were crushed in a mill at Warren to test the value. The three tons yielded $31.10, or that is the amount that was taken from the plates. The battery was not cleaned. A big mill will some day be put up on Thunder mountain which will in a not very distant day no doubt be one of the greatest camps in the state. Citizens of Boise are now talking of constructing a wagon road to Thunder mountain. The cost will be about $15,000. The Caswell Brothers have a team and expect to drive as far as Penn basin, 25

miles from Thunder mountain. The route is by way of Bear valley, where they leave the state road. From there to Penn basin there is almost a natural road.

But, aside from a passable road through Penn Basin, more was needed than some small, yet profitable, surface placering of an outcrop which could not be identified as a vein. They had a rhyolite cliff with values exposed to a depth of two hundred feet along a length of five hundred feet. This whole surface ran an average of $14 a ton. Some strata went as high as $100 a ton, and some pans as high as $10. William E. L'Hame asserted: "It is impossible to pick a sample on the face of this cliff that will not pan, and yields of fifty cents to one dollar to the pan are quite frequent." Volcanic basalt and other intrusions made for complicated geology. Yet L'Hame noticed that "mineralogically and structurally it is strongly suggestive of the famous Cripple Creek region of Colorado." Because recovery so far at Thunder Mountain had not exceeded half of what gold was processed, L'Hame foresaw a truly bright future. So was it reported about W.H. Dewey:

> He [Dewey] immediately put men to work developing them and several tunnels were run a distance of eighty-four-feet and the ore taken out went $13.50 to the ton. In another much shorter tunnel the ore went $8 to the ton. Up to this time, not including that taken from the tunnels, 460,000 tons of ore has been broken. Mr. Reed, an experienced miner, who has been at work in these mines, has reported that he fully expected to find a sixty foot vein.

> Colonel Dewey has ordered two mills from Chicago, one of which will be here by the 10th of June. One is a 10-stamp mill, which will crush from 50 to 70 tons in a day, the other is a 100 stamp mill and will crush from 500 to 700 tons of ore. The wagon road at present only goes to within 50 miles of the mines and will not be built in this year, as it is thought everything needed can be packed in. The colonel thinks he now has without exception the best mines in the United States and he is going to push the work of development, and it will not astonish anyone who knows anything about this region if the greatest gold discovery of the age is made here.

Impressed with the results of his investigation in 1900, and with Thomas C. Reed's report in June 1901, Dewey decided to go ahead with the purchase. Reed had shown Thunder Mountain to F.J. Conroy of Pittsburgh, who

> brought out 100 samples for assay. Each represents 100 pounds of ore, carefully sampled. Each 100 pounds was crushed, mixed and quartered. The samples were taken from the tunnel, cross-cuts, croppings, and slides. Samples were taken every five feet in the tunnel and cross-

cuts . . . . Samples were not taken from all the claims, and this was not really necessary. A sample of one is virtually a sample of all. Assays will be made of samples taken from three or four of the claims.

Dewey sent out eight miners with Locke's pack train. This will make a force of eleven men. Three miners have been at work this Spring and Summer. Four were employed during the winter. The tunnel, with cross-cuts run each way, measures about 300 feet.

At this stage, the Caswells professed disappointment. With additional help from William Huntley, they had increased their total recovery to over $20,000. While so engaged, they had come across a truly rich pay streak that they thoughtfully covered up. A five- to seven-foot vein, forty feet long, ran as high as $9.80 a pan. An especially valuable three-foot section of the vein yielded $3,000 a ton. "Four sacks of ore taken from a width of seven feet gave returns of $1997.21. Another sample taken where the ledge is five feet wide, assayed $1000.83." Sometimes they contemplated developing the property with Richey and Huntley, hoping that Dewey would reject their offer to sell. But Dewey's energetic development dashed any such hope. (Actually, they were just as well off to realize $100,000 instead of losing a fortune that they did not have.) So they finally chose to get out as soon as possible. Using their rich new find as an incentive, they induced Dewey to complete his $100,000 purchase immediately, rather than wait until 1902 as originally contemplated. Dewey was going ahead anyway, so their insistence did not inconvenience him much just then.

Very flattering reports continued to emanate from Thunder Mountain even though those better discoveries remained confidential. A prominent, but unidentified, Boise miner told E.W. Johnson on July 9, 1901:

After thoroughly looking over this country I am forced to believe that the greatest mining camp in Idaho will soon spring up in the Thunder Mountain region. All it will require is a little nerve and capital to open up these mammoth mineral veins. Prospectors are coming in from all sides and already vacant ground is scarce. Several parties are here waiting for their associates who are coming in with money to secure property and there is no doubt but there will be lots of development work going on here soon. Every day there are reports of rich strikes being made all around here and these reports are backed up by samples of the ores found. Never before have I seen such surface showings as in this camp. With capital to open these big ledges Thunder mountain would astonish the mining world.

Thunder Mountain mining area

Information reaching the *Lewiston Tribune* from Grangeville was equally enthusiastic:

Parties arriving from the Thunder Mountain country report that the work has been progressing all winter on the property of Col. Dewey and has shown up an extent of ore and values that is marvelous. This has always been considered one of the richest sections in any part of the State, and indeed ever struck. When the property was in the hands of Dewey's grantors and Mr. Woodruff was managing the mine, he assured his company that if they sold to Dewey he and not they would make millions out of it. But they sold and the facts so far demonstrated proved him right. The property under the management of Dewey has been worked all winter by a force of from eight to twenty-five men, and a large number of cross-cuts, tunnels and shafts have been sunk to ascertain the extent and values of the ore body. It is demonstrated, as has always been contended, that the whole mountain is practically one dyke of ore. Assays have been made nearly every day of a large number of samples from all the workings and the company claims that, making every allowance that could be asked in computing the average values, and then cutting the result in half, the ore in sight if it goes no deeper than the shafts so far sank, which is about 110 feet, is worth $6,500,000. Assays are taken to average about $6 to $8 for the whole body of ore. The enormous total is practically only the surface of eight claims. Other properties in the region are as good apparently.

S.B. Edwards, "one of the best known prospectors of Idaho," had encouraging information based upon his own experience there in 1899 and 1900.

A year ago this summer he [Edwards] discovered some float which struck him as representing something of value. Upon his return in the fall he had assays made of this float both by James A. Pack and Thos. Manning. Pack's assay gave values of $48 in gold and 6 ounces in silver. Those of Assayer Manning were $44 in gold and 4 ounces in silver. These most encouraging assays of the float led Mr. Edwards to return to the Thunder Mountain country early this spring and seek for the ledge from which the float had come. His efforts led to the discovery of a blind lode upon which he has made three locations all showing ore of a most excellent character.

Ores from these claims have also been assayed by Albert B. Sandford, assayer of the custom house in Denver, Colorado. Mr. Sandford's assays show considerable higher value than those of Pack and Manning, some of them going as high as $100 in gold and 48 ounces in silver and from 60 to 70 per cent or nearly $600 a ton.

He further declares that there is room for a thousand prospectors yet remaining. The country is not half prospected. There are probably at this time in Thunder Mountain and the section surrounding it 400 men, but new prospectors are coming in every day. These come from all quarters, from Warrens on the west, Salmon City on the east and from Boise and the southern section. The number now in there will be doubled before snow flies.

Mr. Edwards says he looks for the biggest finds to be made in Thunder Mountain that the northwest has ever seen. He has unbounded confidence in that section and says that Thunder Mountain means to Idaho what Cripple Creek and Leadville mean to Colorado and that the district is a far more extensive one than any that state ever saw. Thunder Mountain itself is nothing but a mass of ore. This has been fully demonstrated by the operations of Col. Dewey and his associates. It is ore everywhere. To be sure much of it is low grade, but the almost limitless quantities in sight will make it one of the most productive sections of the world. He prophesies that it will more than equal the celebrated ore mountain of Treadwell's Island in British Columbia.

⚒   🐴   ⚒

A genuine Thunder Mountain mania was finally built up from the impact of countless reports of great mineral wealth. Thunder Mountain had a romantic name anyway: acting as a sounding board for lightning which danced off nearby Lightning Peak, that somewhat inconspicuous mountain offered legend writers a welcome opportunity to display their talents. While Thunder Mountain was gaining interest everywhere, some practical problems had to be faced. W.H. Dewey raised enough capital in Pittsburgh to assure purchase and development of his Caswell property. To get a road necessary for hauling in a large plant, he offered to put up $10,000 of a $20,000 estimated cost if Boise subscribers would match his share. While they

contemplated this venture, Dewey's initial ten-stamp mill — designed so that packers could get it in over a mountain trail — arrived for shipment to Thunder Mountain. Upon a very strong positive engineering recommendation, Boise's Chamber of Commerce decided on August 16, 1901, to join Dewey's road project. John Pilmer, their agent, assured them that

Thunder mountain and its vicinity was an entire revelation to him. Before going in on this trip, he was considerably prejudiced against that section but he returns firmly convinced that Thunder mountain is the greatest mining camp on earth today. He says that the entire mountain is a solid body of ore. In appearance it is very similar to some of the white, chalklike cliffs of the Snake river.

The mountain has a topping of lava, which is broken away upon all sides. The formation itself is a very soft porphyry, every part of which is ore. This decomposes very rapidly upon exposure and from this source came the gold discovery by the Caswell brothers. The whole side of the mountain disintegrates by the action of the elements and sloughs away to the lower ground. This has been washed in the placer mine with excellent results but the dumps are still nothing but quartz containing much gold.

The Dewey tunnel and all those being driven into the mountain are all in ore. There are no hanging or side walls and but little or no difference in the value of the ore at any given point.

Mr. Pilmer thinks that open cut mining, similar to the operations for many years carried out at the famous Treadwell mine in British Columbia, must be followed at Thunder Mountain. He says it is the greatest proposition he ever saw and states that a 1000-stamp mill, if started tomorrow and operated continually for a thousand years would still leave vast quantities of ore unmilled.

Reassured by this enthusiastic response, and in need for early construction before another winter's delay intervened, Dewey offered to go ahead, advancing the initial costs. Then when Boise subscribers succeeded in raising only $3,000 (less than a third of their share), Dewey became impatient. After more than a month's delay, he called his entire proposition off and decided to build his Dewey Palace Hotel in Nampa and to find a Long Valley route for his road. Meanwhile, he had his original mill packed in through Bear Valley and Penn Basin along his original route. This involved great expense. Lem York reported:

Supplies and machinery are freighted to Bear Valley, about 100 miles above Boise, where they are transferred to the big pack trains and transported 80 miles further to the mines. To one who has not been over the route no conception of the

View of Thunder Mountain

difficulties encountered can be had. Every pound of freight has cost the operators 6¼ cents per pound, or $125 per ton, and it is safe to say that the transportation charges have greatly exceeded the first cost of the invoices.

At the date of our visit, September 1st, the camp presented a very lively scene. Men were hurrying here and there: trails and roads were being graded; wood and timber for the mill was being 'snaked' in from the surrounding timber; carpenters were busy erecting a two-story boarding and bunk house; the mill grade was ready for the foundations and most of the machinery was piled in the yard ready for erection. Prospectors, with their pack outfits, attracted thither by the stories of vast wealth, were coming and going, quite a 'tent town' being established on the bench above the mine. It was a scene calculated to take the mind back to the pioneer days of Idaho, for Thunder Mountain certainly occupies a frontier position.

Through the courtesy of Supt. Reed, our party was taken through the now famous mine, which has been opened by cross-cuts and drifts aggregating some 550 feet; all in ore. The work has been done in the shape of a cross, thoroughly demonstrating the uniform value of the rock. No timber is required. The ground is easy drilling and breaks fairly well. Fifty-two samples, each weighing 200 pounds were taken from the property, outside and inside, a few weeks ago, the average value being a fraction over $6 per ton in gold. When it is considered that this ore can be mined and milled for less than $1.00 per ton, its value can be partially realized. It is calculated that seventy tons can be treated daily with a

10-stamp mill now in course of construction. The ore is very free milling, the values being readily saved on plates. Some twenty-five men were employed in and about the property at the time of our visit, but we understand that the number has now reached about 50.

Considering Thunder Mountain's potential, this effort was worthwhile:

Its formation (in the language of the prospector) is porphyry and basalt, the line of contact extending nearly east and west, the south side of the mountain being basalt and the north side porphyry. And it is a mountain of gold! Whatever its origin — whether it came up, fell down or slid in, we cannot say — but the fact remains that the whole mass of conglomerated material carries the royal metal in paying quantities. There are no veins. Not a piece of quartz, even, can be found in that marvelous monument of mineralogy.

Martin Curran, who completed Caswell sale arrangements, returned with assurances that his $100,000 investment was purchasing two million tons of ore. Having been shown the secret Caswell discovery, Curran had interesting statistics to support his optimism:

Nature did wonders for this property, as the great vein or zone stands up from one hundred feet on the westerly end to one hundred and fifty feet on the easterly end, over the level of the surface exposing the great ore body for more than three thousand feet in length, and from one hundred to one hundred and fifty feet in width, leaving exposed one million five hundred thousand tons of pay ore at a conservative

Thunder Mountain tug-of-war and sack race shortly after 1900

estimate of ten million dollars.

The underground workings consist of about five hundred feet of cross cuts and drifts, every foot in pay ore. Main cross cut, sixty feet, samples seven dollars and eighteen cents, pay ore still in face west drift cross cut fifty-five feet, six dollars and twenty-seven cents. Face of west drift, seven dollars and eighty-two cents dark ore. On south side of west drift one hundred and forty-eight dollars and twenty-nine cents. At this point it requires a cross cut south to determine width of this high grade ore, also forty feet cross cut north to go through ore, such as the face of the west drift. At present it is unnecessary to do any more work in the mine until the mill starts, as it is easier to handle the ore from the mine than the dump.

On the surface and about the center of the great ore body and between the two underground cross cuts there is a very rich chute of ore, forty feet long and from five to seven feet wide, that assays as follows: Seven feet, $1975.84, $1000.93; five feet, $199.78, $266.20. If this rich chute carries the same values to the tunnel level, same width and length (estimate one thousand tons, average value $860 per ton, or $860,000.00), the property can furnish one thousand tons of ore per day, as soon as there are a few chutes put in the mine to load cars from, and can be mined for 60 cents per ton. The property requires 200 stamps, and with that number in operation the property will pay $150,000 per month.

The property is situated near plenty of wood and water and can be worked by tunnels for a great number of years. The saw mill is all on the ground and will be sawing lumber by the 10th of the present month, also ten stamp mill in course of erection and will be in full operation about December 1st, 1901.

William E. L'Hame concurred:

I consider the formation identical with that of Cripple Creek. It consists of royalite intersected by phonolitic intrusions. The greatest values are met with at the contact of the dyke with the overlying volcanic breccia. The position of the dyke shows that it was one of the last of a series of volcanic actions which took place at a period probably antedating the Cambrian age. At the intersection of the dyke with the other strata the same is crumpled and crushed, giving special opportunities for the mineralization of the same.

It is also proper to assert that the carbonaceous material which is found in the volcanic tufa in the form of fossil has the effect of precipitating the metal from the auriferous solutions which accompanied the dyke during the process of eruption.

I believe the ore presents special facilities for free milling on account of its chemical composition and makeups, pyrites and other base metals being almost entirely absent. Samples taken from the deposit showed values of nearly two thousand dollars per ton, ranging downward, too, of course, less in places. It shows on the surface several hundred feet of valuable ore that has been exposed by hydraulic workings a distance of maybe five or six hundred feet in length, and two or three hundred feet wide. The values so exposed will probably range all the way, as far as I am able to say, from two to three hundred dollars to as many thousands. Mr. Richie showed me a place about as large as this little corner by the door in which they took out three thousand dollars. Mr. Richie panned $2 from one pan of dirt he took from the top of the hill. There is an immense amount of slide rocks and the dirt between it all assays very big. There is a great amount of gold in it, and all the tailings that have gone through the sluicings contain a very appreciable amount of gold. There is an immense amount of ore — a whole world of it. I think all the slide rock has gold in it, and if it is all auriferous material there is a million tons of ore

in sight there. That is a very fine mineral section
in there.

Thunder Mountain is a mountain of ore;
there is nothing like it; it is no hill, it is a mountain. There is all the reason in the world to
believe the deposit is continuous. I have not seen
any mine in the country that makes as fine as
Thunder Mountain. I consider it second to none
in the United States.

E.W. Burton of Murray came out with a slightly
different impression of Thunder Mountain.

[Burton] formed a favorable opinion of the new
camp, and declares it to be unlike anything ever
discovered. He claims that it is simply one vast
field of decomposed mineralized rock. Some call
it quartz; others designated it as porphyry. Its
real value as a whole had not been determined,
as there were miles of it. It looked like a vast
overflow of some crater, which spread as it continued to discharge. There were no doubt some
rich streaks in it, but the mass was low grade,
and the methods of working would have to be on
a very large scale, so that thousands of tons could
be reduced every 24 hours. Water is plentiful
within a short distance, but fuel is scarce. It is
not, in a strict sense, a poor man's camp, but
many miners will be employed there in time.

An even wilder report emerged from Challis:

Thunder Mountain is all the rage in this part of
the world. There is nothing peculiar about
Thunder Mountain to look at in the distance. But
when one gets to it there is something peculiar
about it. Thunder Mountain is a big mountain,
and nearly all the formations that are common to
Idaho are represented except lime. The make-up
of this mountain consists of nearly all the
different granites, porphyry, rhyolite, sandstone
and a little quartz. In the sandstone is found
small seams of lignite stone coal.

While all kinds of accounts of Thunder Mountain circulated during the fall of 1901, Dewey's
crew finished installing a mill engine and boiler
scheduled to start processing ore in December.
Freight costs from Boise ran 12 cents a pound, but
once installed, the mill was expected to process free-
milling ore at only $1 a ton. They also employed a
string of 120 pack mules to get in their winter
supplies late in November. Important new discoveries on Monumental Creek three miles from
Dewey's property added a new dimension to the
mining, and accounts of other valuable finds over a
broader area created still more excitement.

In preparation for a grand rush, Boise's
Chamber of Commerce went ahead in constructing
bridges and road segments which W.H. Dewey had
backed off. The cities of Weiser, Emmett (rail

Building a cabin at Thunder Mountain

terminal for Dewey's new route), Grangeville, Dixie, Salmon, Mackay, and Ketchum also began to advertise the routes to Thunder Mountain. Other purchases of claims, mostly selling from $5,000 to $10,000, helped boost interest. W.H. Dewey, returning from Pittsburgh with his $100,000 to complete his Caswell claim transaction on November 16, was reported as being boldly extravagant in announcing his expectations in Chicago:

> Colonel W.H. Dewey of Idaho believes he is the richest man in the world or that he soon will be. There will be trumpet tidings from Idaho within two or three months, he says, tidings that will proclaim Idaho and American Transvaal or a United States Klondike, that will pale the fame of Cripple Creek or any other old diggings. The colonel carries in his pocket a little vaseline bottle filled with pure gold, all extracted from just three pounds of quartz. He knows a man who made a bet that a pound of rock from the new Idaho field would result in from $60 to $80 worth of gold.

Alvin B. and Daniel G. Caswell went out to Ogden and Denver to tell their story of sudden wealth and to explain how they had discovered more mines at Thunder Mountain.

Without waiting to see how his ten-stamp mill worked out, W.H. Dewey ordered a $250,000 second mill with one hundred stamps so that he could process his anticipated $200 million gold mine more quickly. Professor E.H. Mead, while not trying to estimate how much ore could be developed, assured the Union Pacific Railroad that one deposit alone, extending 250 feet in all directions, with no foot or hanging wall anywhere, equalled a 2,000-foot vein. He regarded Thunder Mountain as "the most wonderful mining country I ever expect to see." Already ore "enough shows that will keep a 100 stamp mill going indefinitely." In Minneapolis, Avery C. Moore estimated that fifteen thousand miners would head for Thunder Mountain in 1902 "as soon as the snow melts." Only twelve log cabins and a two-story building were available to accommodate that rush. Late in December, a major stampede was anticipated to Idaho's "Mountain of Gold," in which "Stone Coal, Charcoal, Petrified Wood and All Kinds of Usually Barren Mineral Yield Up Treasures." A Grangeville report on December 26 about L.A. Leland and Frank E. Johnesse outlined Thunder Mountain's attraction:

> 'You may say the truth about Thunder Mountain is fully up to the most extravagant stories that have been published anywhere,' said Mr. Leland. 'It is a geological revelation. The saying is almost universal that *quartz is the mother of gold*. Thunder Mountain disproves this, for there is almost no quartz there, and that little is found only incidentally. But there is almost everything else, and gold in it all. One finds stone coal, charcoal, petrified wood, and all kinds of barren rock here, impregnated with flakes of gold. You can pan gold out of almost everything on Thunder Mountain proper. How it got there, no one knows. A vast primeval conflagration might have melted the gold, and driven it out either molten or in fumes, so that it filled everything. But there it is, a puzzle to all who see it.
>
> 'There is almost no stratification in the camp. It is veritably a mountain of gold. Rhyolite is the chief deposit, although almost everything is found in the most uncommon conglomeration. The mountain has been compared to the Rand reef in the Transvaal.
>
> 'There is probably not a pound of spare food in the camp today. I took in 100,000 pounds of vegetables by pack train from the Caswell ranch, last fall, but I understand they are already short. It would not be advisable to attempt going in now, as one could not take in enough supplies to last, and there is nothing in there.
>
> 'Every foot of ground for the three or four miles square of Thunder Mountain has a claimant. Fractions are eagerly sought. But there may be other districts just as good near by, only they have not yet been uncovered.'

Although more appropriate for a humor column than for an explanation of a $30,000 mine sale, this kind of report was issued all too often as the Thunder Mountain mania built up. Late in December another Pittsburgh sale, this time for $125,000, outclassed Dewey's purchase. By December 28, W.H. Bancroft of the Union Pacific increased his traffic estimate to twenty thousand people headed for Thunder Mountain, with passengers expected from all over the United States.

⚒   🐎   ⚒

As excitement grew nationally, mining at Thunder Mountain slowed down when winter arrived. Dewey's ten-stamp mill was completed on schedule in December, and five stamps were tested. But ore could not be processed until January 3. Winter snow also halted prospecting. Close to a hundred miners worked on their claims, but they could only prepare for later production. Communication was almost cut off with Warren: three miners spent four days, from December 21 to 25, getting out on snowshoes and suffering freezing cold on the difficult trip. Aside from the thirty to forty Dewey employees, who had been taken care of adequately, provisions were scarce and prices were high for all other miners. A November pack train from Warren would have supplied Thunder

Mountain's growing market if snow had not blocked the trail. No more could come through until spring. Warnings were issued to prospective miners to stay away unless they could bring in their own supplies — a practical impossibility. Costs of claims, ranging from $1,000 to $15,000, were also inflated greatly. Fred Holcomb, a pioneer miner, warned:

> Thunder mountain [is] not a poor man's camp, and those who go there expecting to find it one will be disappointed. The whole mountain is covered with place rock, but there is no water that could be diverted to handle it. The Caswells appropriated all the water that could be diverted so as to be used for placering. You see, the placer dirt is on the highest mountain in that part of the country, and the water is all below it. The dirt isn't rich enough to haul it to water.

Another prominent Idaho miner, A.J. McNab of Salmon, although "quite enthusiastic over the possibilities of the country" after prospecting thirty-six square miles at Thunder Mountain, also warned miners in January 1902 to wait:

> It is foolish for those seeking an intelligent foot-hold on Thunder Mountain to undertake a trip into the region at this time as it holds out no refuge to the traveler and exposes him entirely to the chance of reaching a cabin.

Unable to get to Thunder Mountain, impatient miners began to pile up in Warren, ready to dash on in as soon as an opportunity should offer. Stage lines from Union Pacific stations in Ketchum, Mackay, and Red Rock, Montana (operating via Salmon), also prepared in January 1902 to offer service over nonexistent roads (through country in which roads still have not been completed eighty years later) when winter might break. Seventeen Concord coaches were procured for a line from Red Rock alone.

At Salmon Meadows, Charles Campbell (one of Idaho's most prominent ranchers) noticed that local packers were trying to break through to Warren to get in advance of the gold rush. Campbell was reported to have this reaction:

> During his long residence in the west . . . [he had] never seen conservative men lose their heads under pressure as they have done in his section of the state for the past 90 days. 'All the people of the country have the fever,' said the visitor [Campbell], 'and if only one-quarter the stories they tell about the camp materialize, it will be the greatest mineral belt ever discovered. Scarcely a week passes but some one comes out over the trail from Thunder Mountain, and I have yet to meet the first man who says the *find* is not a wonder. When old time miners tell you it is the greatest thing they ever saw, it must have some merit.'

While the townsite of Roosevelt was being promoted in Boise to serve Thunder Mountain, living conditions on the mountain grew critical. When Allan Stonebreaker left Thunder Mountain January 19 on his regular semimonthly mail trip, sacks of flour were selling for $20 to $50 each. Only a few were left. As supplies neared exhaustion, about seventy-five miners had to prepare to retreat from their isolated camp. New rich discoveries, selling for $5,000 a claim, were reported. But right then, groceries were more in demand than gold.

In an effort to alleviate Thunder Mountain's shortage, Frank Andreas set out from Boise on February 3 with a large dog team hauling a ton of provisions. Two other packers left Grangeville with dog teams at the same time. They immediately were followed by an advance wave of the great 1902 gold rush to Thunder Mountain. Packers with horses began to break their way from Dixie through Chamberlain Basin to Thunder Mountain, and other pack outfits were assembling in Bear Valley. Even winter snow could not quite halt the rush. Extravagant testimonials about Thunder Mountain continued to fill Idaho newspapers, and impatient fortune hunters simply could not afford to wait. When Allan Stonebreaker made another postal trip to Dixie on February 6 to 8, he met thirty-five parties of gold seekers. By then the trail was in good shape for horses, with no snow depths exceeding four feet. About everyone who had stayed at Thunder Mountain was selling claims to newcomers. Those who rushed in early in 1902 thus had an unmatched opportunity to assume losses inherent in buying claims there, while those who had spent an expensive winter had a chance to recover some of their investment in time and travel cost.

Winter travel to Thunder Mountain remained somewhat hazardous on some routes. On February 10 at least three miners were lost in a snow slide near Elk Summit between Thunder Mountain and Warren. That misadventure did not slow anyone down, though. Some continued to use dog sleds, although others did their own hauling. A horse and dog sled party left Florence on February 11:

> A snowshoe and rawhide outfit of horses and a large dog train passed through here today en route to Thunder mountain. They expect to be on the road from here about one week. The only trail breaking they will have to do is from the Snowshoe cabin across Salmon river bridge to the Warm Springs, about 18 miles, which they will do with a bunch of horses they have on Salmon river. From the Springs to Warrens and thence to Thunder mountain the road has been kept open all winter. This dog and rawhide train has been on the road from Grangeville to the Salmon river three days and it will take them three days more

to reach Warrens, from which point they will have good traveling. There is plenty of feed for horses along the entire route via Grangeville, Florence and Warrens, with good hotel accommodations.

On February 13, a Grangeville crew included experienced old timers who had been to Thunder Mountain years before, as well as others who had come out that winter:

> A large party of argonauts left Grangeville this morning, for Thunder Mountain. They will go by way of Florence, the state bridge and Warrens, having contracted with Tom Walton to take their supplies as far as Warrens. Each of the men took about 150 pounds of supplies — 50 pounds of flour, 30 of bacon, and the rest of the weight being made up of tea, sugar, beans and bedding. In the party was one 25-35 Winchester rifle, for the deer that are numerous in the Thunder Mountain country; also one 22-calibre for birds and small game. From Warrens, each man will pull his supplies on his own rawhide toboggan.

More dog teams began training in Boise, where problems such as snow slides near Thunder Mountain did not deter those who wished to leave early. Expectations that mills with 2,000 stamps could soon be hauled there made everyone eager to get in while some potential claims were still left to prospect. News of important new discoveries on February 2 arrived in time to encourage more dog trainers. Ben Caswell confirmed reports that he had declined a number of offers of $100,000 for his new claims after selling to Dewey. Finally the Caswells sold their new property for $125,000.

⚒   🐴   ⚒

Caswell estimated from his experience in Michigan and Pennsylvania that thousands of hopeful fortune hunters would soon be headed for Thunder Mountain. Competition between the Northern Pacific and the Union Pacific promotional departments for that traffic grew intense. With a variety of Union Pacific entrances (through Red Rock, Mackay, Ketchum, Boise, Nampa-Emmett, and Weiser-Council) and two Northern Pacific options (Lewiston-Grangeville and Stites-Dixie), local communities throughout Idaho contested for favor. Telephone lines from Boise and Blackfoot via Mackay were also projected to Thunder Mountain.

More snow in March and a shortage of fuel brought additional complications to the miners on Thunder Mountain. After two months of operation, Dewey's stamp mill had to shut down on March 10 for lack of fuel. All timber within a mile had been used up, and additional supplies could not be hauled in then. Everyone wanted to locate more

claims rather than operate a mill. About 250 miners had gotten to camp, but none wanted to work. Trails to Thunder Mountain also became more difficult to negotiate. In February, only two miles were open to anyone aside from hikers who had to drag their supplies over the stretch from Warren. But conditions grew considerably worse. Flour sold at $50 a sack, and food shortages became more troublesome. Boise traffic could get through only by equipping the horses with snow shoes — an old mountain device to facilitate packing:

> The trip from Thunder Mountain to Boise can be made in five days. It took me longer because I stopped on the way. I met two pack trains going in. They were composed of several horses drawing rawhide toboggans. The animals were all heavily loaded. The lead horses of the first pack train wore snowshoes, but the rest seemed to be getting along all right without them. The trains were making about 20 miles a day. I did not know any of the men and did not talk to them. There will probably be a scarcity of horse feed, for the reason that while a horse can draw feed enough to last him along the trail, he cannot do much more, and there is no feed at the other end of the route. The problem of feeding horses will be a serious one before the grass grows in that country.

Late in March, a long train of toboggans and pack horses left Grangeville prepared to break through to Thunder Mountain with almost two tons of supplies. These certainly were needed, as too many gold hunters continued to arrive empty handed.

> 'Not an ounce of food is to be bought in Thunder Mountain at any price,' said Sheriff J. Dixon, who arrived here today [early April] from Warren. 'Men are coming out every day as far as Warren for food, where they can buy staple groceries in limited quantities. It is an 80-mile trip, the way most of the travel goes now, and takes about four days either way. A man who is not well equipped will eat in the eight days coming and going almost all he can carry.
>
> 'Ten cents a pound is being paid for freight from Meadows to Warren. The freighters won't touch it for less. Seven cents is also being paid from Warren to Shaffer's, 25 miles. A party of three Colorado men came through Warren with 1000 pounds of supplies on which they had paid this rate.
>
> 'But the rush continues. One man who came out to Warren last week met four men going in without even a cracker. He divided his last three biscuits among them, telling them that they could buy nothing further on. Still they went in. Others are going not much better prepared. . . .
>
> 'A movement is on foot to shorten the distance over the dangerous Elk creek summit. The wind blows the snow up the long slope from this side,

whipping it over an almost perpendicular descent on the other side, thus forming a comb of snow 50 to 100 feet deep. This is continually breaking off and making dangerous snowslides down into the valley, besides necessitating a long detour. The plan is to dynamite the snow crest, let the resulting avalanche clear a new short road down into the valley and save the long detour.

'Ten degrees below zero was the record all through the mountains last week. Considerable snow has fallen. There are now between five and six feet of snow at Thunder Mountain, according to all reports.'

New and larger mine sales helped compensate for the extended hardship in operations at Thunder Mountain. Another $250,000 Pittsburgh investment, made in frantic haste on March 9, set off a new round of transactions. Marshall Field, S.W. Swift (noted for his meat processing), and George H. Phillips invested $100,000 in fifteen important claims. Another $40,000 transaction infused yet more capital into local mines there. By April, Dewey's company had invested about $700,000 in Thunder Mountain, and that sum by itself amounted to double the total gold production realized there prior to suspension of mining in 1908. Pittsburgh capital also more than made up for Dewey's withdrawal from Boise's Thunder Mountain road project. A total of $20,000 from Pittsburgh was pledged to match Boise's $10,000 goal, so Boise's road builders seemed $10,000 better off than they would have been if they had received Dewey's original offer. Dewey, however, opposed construction through Bear Valley because of the formidable problems his packers had encountered there in hauling his ten-stamp mill. So preliminary planning became embroiled in a hopeless controversy over whether to build and approach from Idaho City-Bear Valley, Placerville-Garden Valley, or Emmett-Garden Valley. Dewey wanted to extend his rail line from Emmett toward Garden Valley, but Boise and Idaho City complained about being bypassed with such a project. Meanwhile, promoters of a half dozen other routes clamored for attention.

Insulated by winter from most exterior anguish over how future miners ought to reach their camp, miners at Thunder Mountain had more than enough excitement of their own. Thomas Johnson reported that "there is lots of ready money in Thunder Mountain. Agents representing all sorts of wealthy clients are there with cash to buy promising properties. The camp is at a fever heat of anticipation." Johnson also had some good stories to tell of skiing in camp and on his trip out to Warren:

'The average Thunder Mountain traveler isn't in it with some of the trained mountaineers who are

going in,' said Tom Johnson, just out from the big camp. 'Why, a fellow got caught in a snowslide the other day, and slid for half a mile, along with rocks, trees and 100 feet of snow. What did he do but pull out his pencil and location blanks, figure out the distances by computing the rate of speed and counting the seconds on his watch, and he located three claims before he reached the bottom. He missed getting the fourth one by just one second. Fact, for I saw the slide.'

Tom came out from Warren on skees, a month ago, and had a notable runaway from greasing the skees with a prepared dope to make them run more smoothly. This time he stuck to plain webs. Dave Pugh, however, who came with him, was on skees, and Tom persuaded him to buy a bottle of the dope. The skees ran away from . . . [him and dumped him into a bank] of snow. Tom says that his lecture on the genealogy of skees was a masterpiece of impassioned oratory, more picturesque even than the grand mountain scenery round about. Dave keeps a bottle of 'slick-em' as a stimulant for his vocabulary, when he runs short of words.

With miners everywhere trying to find new properties, claim jumping became a problem. Snow claims (of the kind Johnson satirized) also became all too popular. Of some 2,800 snow claims filed at Thunder Mountain from December through May, less than fifteen percent had any merit. A stampede to Indian Creek, twenty miles south of Thunder Mountain where prospects had been discovered the previous fall, enlarged this difficulty: "Only snow claims are being staked, the snow being six feet deep."

⚒ 🐴 ⚒

With the approach of spring, a grand rush to Thunder Mountain got under way in April. Impatient prospectors from all over Idaho and many western states filled all available hotels in Boise, Weiser, Pocatello, Blackfoot, Lewiston, and Idaho City, where the Luna House began to look like old times. Early in April, fifteen to twenty were leaving Lewiston daily to get closer to Thunder Mountain. Some—as many as sixty to seventy each day from all sources—were going all the way, although the lack of supplies forced as many to leave, so that Thunder Mountain's winter population (which rose from about 200 to 800) remained at a stable level. Places such as Campbell's Salmon River crossing (between Dixie and Chamberlain Basin) became cities of tents, with 150 to 200 eager miners camped at Campbell's. Organized efforts to open several routes for horses kept packers in Warren, Dixie, and

Bear Valley more than busy. But during April, none could get through. Operators from Warren got three burros across Elk Summit trying to break a horse trail April 18, but more snow defeated their attempt. Before horses managed to get over Elk Summit on May 15, a lone packer had gotten through from Salmon on April 29. Finally on May 12, Frank Andreas of Boise came in from Bear Valley with the first pack train composed only of dogs. His problem, in common with everyone else, had been late snow that was melting, so that travel had become more difficult than ever.

> Dog trains, toboggans and all other modes of taking in supplies give way to packing in on one's back. No one, however, else he may have gone in, comes out for supplies with any other thought than of carrying them in. It is hard work, but is easier, quicker and more satisfactory than any other way. A Boise man who went in with us from Warren, carried two sacks of flour. Another man carried 80 pounds, though most are content to take only 50 or 60 pounds.
>
> We heard of only one man in from Salmon City, though he reported a number along the road. No one has as yet come in from Ketchum and Hailey. Most of the travel is by the Warren route.

Comparisons naturally were made between the obstacles in reaching Thunder Mountain and those of Chilcoot Pass into British Columbia from Alaska that were faced by miners on their way to the Klondike only four years before. A far greater number of men made their way over Chilcoot. But they had to overcome problems which did not begin to compare with Thunder Mountain. The ascent of Elk Summit alone — with close to a 6,000-foot grade — far surpassed anything around Skagway. And that was only one of a number of major hills encountered. Thunder Mountain did not have nearly as severe a supply problem. But difficult terrain for winter travel held back most everyone who aspired to reach Thunder Mountain before spring. Supply trains eventually began to break through from all directions shortly after Frank Andreas' dogs showed what could be done. From then on, energetic miners could reach their destination without undue trouble.

Thunder Mountain received unparalleled publicity during April 1902. Idaho was expected to profit greatly from so much attention. According to Salt Lake's *Mining Review:*

> The tidal wave of prosperity that is about to engulf the Thunder Mountain region in Idaho will bear on its crest many good things for the entire intermountain region. That this wave is coming and will soon be here is indicated in many ways. Every newspaper, at home and abroad, has something to say of this new El

Dorado; of the mineral wealth that has been found within its environments, and of the thousands that will soon be headed toward this promise[d] land, which, from all accounts, will develop a number of splendid bonanzas before the close of the present year. That this will be the case seems almost a certainty, and for two good reasons, one of which is that experts have stated that a vast area of country in this portion of Idaho is heavily mineralized, the other being that a small army of experienced mining men from Colorado, Utah, Montana, Nevada, California and western mining states will thoroughly prospect this region this summer, and if they do not succeed in finding a dozen or more of Monte Cristo propositions, it will be a nine days' wonder. As a matter of fact the whole western country will receive a wonderful impetus, as far as the mining industry is concerned, because of this rush to Thunder mountain, and the entire west will be benefitted. It was a boom similar to the Thunder Mountain excitement that made Cripple Creek, and gave to Colorado one of the greatest gold mining camps upon God's footstool. This movement promises as much for Utah, Nevada, Oregon and adjoining mining stations, within whose boundary lines there are many districts, rich in their deposits of the precious metals, which only need publicity to develop into as great producers as are to be found in the west, and this publicity will naturally be drawn to them as a climax of excitement and attention attending the splendid reports emanating from Idaho's new gold camp.

Hope was also expressed that mining, as distinct from publicity, would help develop Idaho:

> 'Thunder mountain is going to redeem Idaho as a state,' declared Mr. [Frank] Hobbs, 'it is going to make a payroll that will radiate in every direction. The Dewey company now has 100 more stamps going into the mine, and one of the big Pittsburg [sic] companies has ordered 250 stamps for its property on Big Creek, 30 miles from the Dewey. There is none of the usual uncertainty about following the course of a ledge that drops into the earth and no mining is necessary except to quarry out the mineralized conglomerate and convey it to the mill. The life of the camp is not dependent on railway transportation as stamps are the only machinery necessary and they can be carried in by pack train. The camp will make itself without any assistance other than the opportunity it is now receiving so lavishly of being operated in a large and proper way under capable hands.

Along with mines at Thunder Mountain, rival townsites also attracted investment. Not much could be done toward development until spring access became feasible. Roosevelt was projected for habitation all during the incipient 1902 gold rush, and so was Caswell. A somewhat prophetic notice from the *Boise Clipper* raised questions about investment in Roosevelt:

> The town is situated on Monumental creek, and sandwiched between two mountains at an angle of forty-five degrees, and from 3,000 to 5,000 feet high. The creek flows from 600 to 700 inches of water at its lowest, and has a fall of from twenty-five to thirty feet to the mile. The townsite of Roosevelt is about one and a half miles long and from 150 to 300 feet wide. The land is about two feet above low water mark. The stream is noted for its changeable channels caused by snow slides and ice drifts that cause the water to back up until it rises from ten to fifteen feet in places, above low water mark. From Mule up to Coney creek, a distance of about 300 feet, the mountain sloping southeast has been swept of nearly all timber by snowslides which have taken rock and earth and deposited it in the bottom of the gulch where the townsite of Roosevelt is located. The snowslides and high water have destroyed all the timber on the creek bottom as far up as Taylor's cabin. On the northwest slope on the east side of Monumental creek the snowslides have not done so much damage, as the mountains are covered with a dense growth of heavy timber. But, if Roosevelt should make a town this timber will be used and then there will be no protection whatever against snowslides and the danger will be greater, and some day Roosevelt will be wiped from the face of the earth. This year is an exception, there being only from five to seven feet of snow, and the sun hardly ever touches the snow on this slope.
>
> If a man should build his house on stilts out of the way of high water, a snow slide is liable to come along and knock the props out from under him, and if he protects himself against snow slides the high water will drown him. Continual displacement of about five feet a year created havoc by misaligning placer ditches.

Although more optimistic observers scoffed that no avalanches disrupted life there in 1902, a potential problem still remained.

When melting snow finally allowed impatient packers to break some final barricades which had obstructed passage to Thunder Mountain, the great 1902 gold rush finally surged into Idaho's most remote mining camp. Considerable effort was required to surmount high drifts blocking ridges such as Elk Summit:

> Three hundred and fifty loaded horses, and 100 men, crossed Elk Creek summit into Thunder Mountain, Sunday, May 25, is the word given out by O.H. Benson, of Florence, who came out from that camp yesterday to bring the good news.
>
> An army of men shoveled snow all Friday and Saturday, and opened the trail. They are pouring into Thunder Mountain now, a regular

cataract of men. Everybody goes in by way of Florence, or up the Salmon, through Warren, these being the only routes open. Thunder Mountain is fairly flooded with supplies.

Within a week, another 1,200 pack mules and horses were lined up ascending Elk Summit bound for Thunder Mountain. The completion of a Salmon River ferry helped a throng of miners and packers get between Florence and Warren on their way to Elk Summit. W.H.V. Richards, a Thunder Mountain pioneer, who came out while most people were headed in, told a reporter how the ferry service helped out:

> About 50 men and 425 horses were waiting to be put across the day it was completed. The road is lined with those going to the camp and old Alaska travelers say the crossing of the Elk divide seems like a miniature of the Chilcot pass in the numbers of men and horses which are constantly pushing over. When the road was first shoveled through two weeks ago about 500 men were camped on this side and pushed on over. Mr. Richards estimated that on his way out he saw from 1000 to 1500 men on their way in, either on the road or camped along the way. He says there are probably 4000 people now in the district, and before August he looks for 20,000 people to be there. Arrivals are coming in constantly increasing numbers by Salmon City, Council, Grangeville and Dixie routes, but the majority are going in by way of Warren. He came out by Warren and Florence and says that the only snow now to be encountered is patches between Adams and Florence; that the ground is shoveled bare on the Elk divide and that at the camp itself there is fine bunch grass and good feed all the way. Mr. Richards says that the Cripple Creek men who are at the camp (and the whole world has its representatives there) are enthusiastic over the prospects and say there will be at least 8000 people come to the camp from Colorado alone.
>
> There is a town of about 100 tents on Marble creek, and log houses are just going up. On the west fork of Monumental creek; about two miles from Roosevelt, is another town site which is controlled by the O. R. & N. railway people and is most probably the coming town as there is a 200 acre flat which affords room for building, while the other sites are too narrow and steep in the canyons to permit much of a town being built.
>
> There are three stores and supplies are plentiful at fairly reasonable prices. Flour is $10 per sack. Some supplies are coming in by way of Salmon City, and pack trains are scattered all along the Warren road with provisions.

More than a few unusual outfits came in. One miner approaching Thunder Mountain through Idaho City pushed his belongings in a wheelbarrow. Another packer used cows instead of horses

Cow train to Thunder Mountain in 1903

or mules:

> Everybody the past two or three days has been
> anxiously awaiting the arrival of the cow train
> from Boise. News of their approach was
> telephoned from the office of the Dredge
> company. Men and women — young, old, and
> middle-aged — as well as children, were out on
> the streets in groups, awaiting the coming of the
> caravan. The people here have seen almost all
> kinds of trains, and cows are not a curiosity, but
> a cow train packed with provisions, camp outfit
> and other things, is a novelty not often seen. Of
> the nine animals, six cows and a red bull were
> packed. They jogged along with their loads as
> gently, leisurely and contentedly as if they had
> followed the business from calfhood. The owners
> of the train are Homer I. and A.D. Clark. The
> wife of the former is with the train, and was on
> horseback, with a child in her arms. They are on
> the way to Thunder.

When a small dairy operated all summer supplied
by pack cows which had resumed their normal
occupation, the superiority of this transportation
system was demonstrated effectively.

Although estimates of 4,000 miners were made
by optimistic promoters, about 1,400 actually
reached Thunder Mountain early in June.
Eventually around 2,000 may have gotten there in
1902. Fortunately all but about ten percent of some
20,000 expected gold hunters stayed home. If they
had not, Thunder Mountain would have faced
serious problems compounded by isolation and
difficult access. At best, this district offered little or
nothing to an ordinary prospector. Thunder
Mountain was a rich man's camp — not a poor
man's. Most of the 2,000 to 4,000 expectant miners
who actually got there were poor men who served
no function in the area. Some went out to find other
mining possibilities in nearby districts, but most
simply had to return home.

Large investors made out better than
impecunious prospectors did — at least until they
found that they had invested in unproductive

mines. Some of them exercised appropriate caution
in getting expert evaluation of potential mines, and
still went wrong. H.L. Hollister of Chicago came
back at great expense with a party of twenty horses
and ten expert appraisers and engineers. He also
had William Allen White, a noted Kansas author,
along to enjoy the trip. Hollister acquired a number
of important properties in nearby districts as well as
the Thunder Mountain district after verifying
values which would "warrant me in giving support
to the district. It is a low grade proposition, but a
place for big men with large capital, and for big
mitts with large capacity."

Lewis N. Clark, sent out by James Guffey of
Pittsburgh to supervise construction of a Trade
Dollar power dam at Swan Falls, also investigated
Thunder Mountain properties to justify an addi-
tional $100,000 purchase by Pittsburgh investors.
This action raised Pittsburgh capital at Thunder
Mountain to about $1 million.

Other observers concurred. A Thunder Mountain correspondent of Portland's *Morning Oregonian* explained:

> I have carefully weighed all the evidence for and
> against it, have prospected the rock, had it
> assayed, and seen as much of the country as was
> to be seen. I unhesitatingly say that up to the
> present time there is not a particle of evidence
> against the camp, absolutely none; and there is
> much in its favor.
>
> The development of the Dewey group has
> thus far shown it good, and with every
> probability of its being a big property. Back of
> Thunder Mountain is Lightning Peak. Some men
> brought down surface dirt from it and rocked out
> over $50 in a couple of hours. South of the
> Dewey, and following its general strike, values
> have been found all along. From the rock, from
> an assessment hole, a mile to the south, I got an
> assay of $5.29. The owner assured me it would
> not carry anything, and the reason he had done
> his work there was because it was the easiest
> place where it could be done.
>
> Across Monumental, and to the west of
> Thunder Mountain, porphyry dykes bearing
> close resemblance to that of the Dewey cut the
> mountains with the same general direction or
> strike. Along these dykes good values have been
> found, and reliable and disinterested men have
> assured me that they have found colors in
> panning over a considerable area of that country.
> The same can be said of the Sunnyside district.
> Surface values can be found in every direction.
> That they will go down development alone can
> tell, but it is far from a discouraging sign of value
> at depth to find value at the surface.
>
> The important thing for the prospector to
> learn in the Thunder Mountain district is what
> rocks carry the values. He will find different

conditions than he has probably encountered elsewhere. There are no quartz ledges in the immediate vicinity. The value seems to be in the porphyry. Rock of this character, that I brought out, and which had 'a lean and hungry look,' and I would have pronounced valueless, assayed well. A creamy white porphyry carrying large white crystals of feldspar went $12.85; a blackish blue rock yielded $29.68 gold. Of course there is ore of much higher grade. I brought out a slab of rock as large as my two hands that is plastered with gold — it is a specimen, and a pretty one, but not to be taken into account in reckoning the camp's possibilities from a business standpoint.

Some disagreed:

The general impression among some very conservative mining men now in the district is that there is nothing whatever so far developed to justify the boom; that it is the most overestimated district that has ever been foistered on the public, and that there will be quite a string of disappointed investors, who paid fancy prices and forfeit money, when they have had a chance to examine their claims.

There has been nothing of definite value developing on any group of claims in the district so far, outside of the Dewey group, and it is considered by some that the Dewey itself has not developed a pay-ore capacity in excess of its present equipment, which consists of a 10-stamp mill.

But major claim sales continued. W.E. Pierce of Boise realized $40,000 selling Thunder Mountain properties in the east. F.W. Holcomb of Salmon did better on a $65,000 transaction with Thunder Mountain's largest purchasers. Boise attorneys, James H. Hawley and W.H. Puckett, made $73,000 in a Philadelphia transaction, and late in June, another $100,000 purchase came from New York. Smaller, yet important, sales continued to help support mining speculation at Thunder Mountain.

※   🐎   ※

Townsite development also absorbed Thunder Mountain capital while gold fever ran unabated. Only two saloons and three stores served all of Thunder Mountain early in May. With a great influx of miners, Roosevelt alone gained thirty-seven saloons out of 150 licensed (but not necessarily operating) for Thunder Mountain. By July, Roosevelt, Marble City, and Thunder Mountain City had about 400 population, and another 1,200 were out prospecting in an area of ten square miles. Two other cities, Caswell (an unsuccessful promotion) and Copper Camp, had less to offer. Fourteen saloons, ten stores, two butcher shops, two drug stores, a restaurant, and a barbershop

survived in Roosevelt until mid-July. All these enterprises, along with a residential district were accommodated in forty-two tents and four log structures. By that time, Thunder Mountain City had ten cabins and forty tents to house 250 miners. A month later, Roosevelt still had ten saloons, along with three dance halls to entertain miners in their leisure hours. A population of about 2,000 remained around Thunder Mountain in August. Five major mines employed fifty, forty, twenty-five, twenty, and fifteen men each. A number of smaller mines continued to operate, but most gold hunters preferred prospecting to working in mines.

In contrast to some other western mining camps, Thunder Mountain escaped an era of crime and disorder. Some claim jumping problems created concern. Thunder Mountain, in fact, never did have to dispose of criminal cases. Mining values — or the lack of them — did not attract a criminal element. Only one mill accounted for all production of any consequence, so little of any value was around to tempt a potential robber.

The high cost of living, amounting to $5 a day into July, continued to restrict companies from hiring miners even if they could have found any willing to work. Even Dewey's mill had to shut down until prices declined in mid-July. Then this pioneer mine and mill managed to operate three shifts a day developing a large block of $7-a-ton ore. A crew of thirty to forty men identified an orebody 2,000 feet long, 140 feet wide, and 180 feet deep. Ore taken out during development kept Dewey's ten-stamp mill busy day and night.

Several other companies devoted the year, 1902, to essential development work. They certainly needed to. One had invested $125,000 in a Caswell property that had been tested to a depth of only ten feet. Generally the companies reported encouraging results, although some perceptive miners anticipated the trouble in risking so much acquisition and development expense on untested properties.

Two major companies decided to ship their one hundred-stamp mills to Thunder Mountain in 1902. One came on thirty freight cars from Thomasville, North Carolina. Dewey's mill filled forty freight cars from Chicago. Both trains had advertising streamers announcing that they were headed for Thunder Mountain. Dewey's mill got as far as Emmett. Lack of a road forced him to park it there for the rest of 1902. Boise's road got as far as Penn Basin that fall, but that did not help.

Dewey still had great confidence in his venture in spite of disappointing delay. He still was looking for miners to work all winter.

'If we can get them,' says the colonel, [Dewey] 'we will keep employed in the neighborhood of

Thunder Mountain miners

50 men. Superintendent Frederic Irwin's reports are very gratifying to us, and we have every confidence in our Thunder Mountain mines. Why, if the assays show only $4.50 a ton with the inexhaustible ore body, we will add 500 more stamps, but I am convinced from the returns we have had that the greater portion of it will go much higher. We expect to spend $1,000,000 before we get anything in return from these mines, and we are making this outlay with the utmost confidence in the district.'

⚒   🐴   ⚒

Plans to retain around 600 men at Thunder Mountain for a winter season did not materialize during 1902. (Even if this arrangement had worked out, some 400 to 500 would have had to leave for lack of provisions.) Roads from Bear Valley, Warren, and Yellow Jacket (giving access to Salmon) were stocked for winter travel. But an extremely heavy and unexpectedly early three-week snowfall in November cut off Thunder Mountain before winter supplies were in. Only 240 miners could be accommodated, so development was restricted.

Legislative support for a Thunder Mountain road led to funding of a state wagon road from Long Valley to Roosevelt. Two season's work went into construction, so both of Thunder Mountain's large one hundred-stamp mills had to wait until September 1904 before transportation became available to Roosevelt. Development work continued on three significant properties, employing forty, twenty-five, and twenty miners. One small five-man operation and about fifty contractors made up Thunder Mountain's labor force for 1903. Enough was accomplished to demonstrate to George H. Williams, an experienced Idaho mine evaluater, that "there are no mountains of gold there," but that Thunder Mountain "will develop into a good camp." One property had two thousand feet of tunnels and shafts; while another had one thousand feet. With Dewey's property already in production, they showed promise.

Miners' pack train leaving Thunder City

Another purchase of ten claims for $100,000 by Pittsburgh investors and a $10,000 purchase by New York investors maintained optimism in Thunder Mountain.

Four hundred miners were able to remain in camp for another winter, which had fortunately held off until supplies could be brought in. Dewey's mill entered an uninterrupted run of production that returned regular monthly dividends for more than two years from 1904 through 1906. A highly efficient yet modest scale operation made this possible. Monthly production gradually declined from almost $10,000 in 1904 to around $6,900 late in 1906. (A four-month shut down in 1904 for lack of fuel limited that year's total to $78,933.10 from low-grade ore. About $67,000 followed in 1905. In 1906, almost $62,000 was realized in twelve months from 11,784 tons of ore. But production costs of $3.51 on a return of $5.25 a ton allowed for a welcome profit.) However, a ten-stamp mill satisfied all of Dewey's needs, so when a wagon road finally reached Roosevelt, no effort was made to haul in another one hundred-stamp mill which had languished in Emmett for two years. Thunder Mountain's isolation, in this instance, had saved a lot of transportation costs that otherwise would have been wasted.

The completion of a state wagon road enabled 204 freight horses to haul another large mill (reduced, providently, from one hundred to forty stamps) to Roosevelt late in 1904. Winter supplies also could be brought in. But operations did not work out well from that point on. Handicapped by trying to utilize a mill which had been worn out in North Carolina, a large crew of miners (who had done more than two thousand feet of development work the year before) managed to process only 180 tons before the mill broke down on December 21, 1904. Sixty-five miners left camp immediately, and a hundred were gone before winter travel got too bad. Thirty miners tried to resume operations in May, using thirty of the forty stamps which they had available. Then they found that "ore values

disclosed in the extensive development of the mine had been shockingly overestimated and the results produced are reported not to have been sufficient to pay operating costs." Efforts to employ a cyanide process failed, and more milling attempted late in 1906 got nowhere.

Shortages of supplies afflicted Thunder Mountain miners again in 1906, when "life at Roosevelt had few charms." Attempts to improve the situation failed in 1907. Declining ore values held down Dewey's production so much that only a shortened season (April 10 to October 28) could be managed. After 8,920 tons of ore yielded $44,967 in gold, the remaining available rock lacked enough value to pay for processing. After turning out about $350,000 altogether, Thunder Mountain's only productive mine had to shut down. An effort to utilize new rich discoveries failed in 1908. So finally, out of several major mining properties at Thunder Mountain, none, aside from Dewey's modest operation, produced more than $1,000 worth of gold. As a fitting climax to this early phase of mining there, a large landslide on May 30, 1909, blocked Monumental Creek and created Roosevelt Lake. Lasting for two days or so, the slide grew large enough to back up a new lake which flooded the town and forced its evacuation. For the next twenty years or more, buildings floated around in the lake; but as the years went by, they fell apart, and now there are only a lot of boards cast about in the water.

꙼   🐴   ꙼

W.H. Dewey (August 1, 1823-May 9, 1903) never survived long enough to see his Thunder Mountain dream turn into a nightmare. Even his mine, although productive for several years, returned far less than half of his company's investment. Later production, after years of delay, finally increased Thunder Mountain's total to about $500,000. But miners at Thunder Mountain could have thought of easier ways to earn that much wealth. Their experience came entirely too close to matching a Roosevelt miner's misadventure late in 1906. Thawing six sticks of dynamite in an oven proved disastrous to the miner. Blown by the explosion through his cabin roof, he lost his possessions but survived without serious injury. He had a great experience, but had also incurred severe financial loss. Many other miners and investors on Thunder Mountain shared something all too close to his fate.

In spite of failing to match unwarranted expectations, Thunder Mountain had a considerable impact upon Idaho's economy. A flood of prospectors turned up or enlarged important new mining areas. Big Creek, Marshall Lake, and Stibnite all profited from Thunder Mountain activity. More than $1 million sent in from Pittsburgh, augmented with funds from other eastern sources, maintained employment for a substantial Idaho mining camp which produced little aside from out-of-state capital investment. In addition, a considerable share of that outside capital contributed modest fortunes to Idaho claim salesmen, who realized just as much as they would if some of their claims had been worth selling. Many prospective miners had a great, if sometimes disagreeable, adventure.

Eastern investors looked at Thunder Mountain differently. When they made their major Thunder Mountain purchases, they were escaping from other traps which far out-classed Thunder Mountain as errors for capital allocation. When Thunder Mountain fever still ran high, as it did on July 21, 1902, H.E. Taylor explained this situation in detail:

> During the past two years the investing public seems to have been on a continuous investment debauch. Dozens of concerns styling themselves 'bankers and brokers' have sprung up in a day from sources unknown. They have launched all sorts of mushroom industrial, mining and oil companies, which to the close observer plainly bear the earmarks of the 'fake.' A lot of worthless property is bought for two or three thousand dollars and turned over to a company. The 'banker and broker' then spend $10,000 or so for printers' ink and glowing reports of alleged 'experts,' and the public does the rest.
>
> It is the same old story. In times of great prosperity the public seems to completely lose its ability to discriminate between the counterfeit and the genuine in the share market. The operations of these stock highwaymen and financial 'jobbers' constitute the greatest burden legitimate mining enterprises have to bear.
>
> We could have made several deals with brokers of this breed had we been willing to sell them something utterly worthless for a few thousand dollars and furnished them 'reports to order.'
>
> It looks as if a reaction were now setting in, however, [with] the public beginning [to] scrutinize stock offerings more closely. The

> forfeiture of the charters of 269 Texas 'fake' oil companies was a jolt that wakened them up a bit. It is estimated that the 'pirating' oil companies sold nearly $20,000,000 worth of stock during the past year.
>
> Among substantial capitalists and business men there is a healthy and growing sentiment for gold mining investments [sic]. The favor so long enjoyed by 'coppers' had been killed and the speculative manipulations of the insiders of the Amalgamated trust and the resulting decline of 20 to 50 per cent in nearly all the leading copper stocks and in the metal itself. New England, the home of the red metal shares, had been hard hit, and is now in a state of investment apathy.
>
> New York and Pennsylvania escaped with only slight injuries and are practically confining their mining investments to gold properties.

Naturally, Taylor did not intend to have Thunder Mountain provide a similar deception for unwary eastern investors. Some mine appraisers and mining engineers were prepared to show caution in 1902. But too many were carried away in a speculative mania. So Thunder Mountain was misjudged and over-sold in a manner reminiscent of fake oil companies and copper mines whose dealings he did not want to emulate.

Thunder Mountain's early production of nearly $350,000 prior to 1909 finally was raised to about $500,000. Later operations from 1937 to 1941 account for almost all that addition. Then after an almost forty year lapse (aside from a minor yield in 1946), activity resumed on Dewey's old mine in 1980-1981.

Gold rush episodes comparable to Buffalo Hump and Thunder Mountain occurred somewhat rarely. Yet they can be attributed to an economic arrangement which featured profitable ventures along with gross errors. Excessive costs associated with improvident examples such as Thunder Mountain ought to be assigned to an overall balance which accounts for successes as well as failures. Information gained from colossal disasters such as Thunder Mountain constituted part of a mining heritage which continued to develop western mineral wealth on an enlarged scale for subsequent generations.

Part V. Afterword

A half dozen mining areas arose in southern Idaho from twentieth century prospecting, although two of them had been anticipated earlier. Some of these districts are of major importance, while others have not produced very much.

STIBNITE

Surpassed only by the Coeur d'Alene, Boise Basin, and Wood River mining regions, this camp finally assumed some of the importance that originally had been anticipated for nearby Thunder Mountain. Discovered during the Thunder Mountain rush, Stibnite developed slowly because of its isolation. Gold and antimony claims were recorded there in 1914, and a mercury excitement (encouraged by a wartime shortage) occurred in 1918. (Another wartime mercury shortage helped to make the Stibnite area the second largest producer in the United States in 1943.) After F. W. Bradley acquired the mines in 1927, full-scale development got underway, and the production of gold and antimony began in 1932. Important tungsten deposits came into production in 1944, and during the war Stibnite was the leading tungsten producer in the United States. Total yields for the active period, 1932-1952, amounted to $24 million in antimony, $21 million in tungsten, $4 million in gold, $3 million in mercury, and $1 million in silver.

Improved gold and silver prices induced Canadian Superior to install a pilot plant at Stibnite to test ore in 1978. By 1980, $10 million had been invested in developing a major new operation that only required the completion of an environmental impact statement in order to get clearance for production.

PATTERSON

Patterson Creek was prospected for gold and silver as early as 1880, but the veins proved to be too low grade to work. After the discovery of tungsten there in 1903, production began in 1911, and a mill was constructed in 1912. Limited production followed during the war, but major development did not come until 1934. Activity continued until 1958, when the mine shut down and the equipment was sold. Production by then had reached $10 million.

LEADORE

Prospects in the Leadore region were known earlier, but the Leadville mine was not located until 1904 and not productive until 1908. Construction of the Gilmore and Pittsburgh railway solved the transportation problem, and as early as 1912, $100,000 had been realized. Nearly $300,000 finally came from the district.

PARKER MOUNTAIN

Discovered in 1904 by Challis prospectors, this upper Warm Springs Creek district had an eight-foot vein of $30 ore, with gold and silver assays as high as $2,365 a ton. With fifty claims, this fairly small district, eventually drained by Loon Creek, generated considerable excitement for a year or more. After showing more initial promise than later production, Parker Mountain finally accounted for $10,000 by 1915.

WEISER MINING AREAS

Several mining districts in the Weiser drainage, with at least one district known not long after 1890, gained early twentieth century attention. Claims recorded on Monroe Creek (1907-1910) and Hornet Creek (1908-1910) provided mining interest for several years. Other areas around Weiser were somewhat conspicuous as well.

More important was a large mercury mine near Weiser. Discovered in 1927, this property was leased, after additional finds in 1936, to L. K. Requa, who formed the Idaho Almaden Company; between 1939 and 1942, $750,000 (4,000 flasks) was recovered. Another leasee brought the largest rotary kiln installation in the United States and began to produce two hundred flasks monthly in September 1955. An additional $1 million of pozzolan has been produced from the calcined mercury tailings. This raised the Weiser total to about $8 million by 1969.

BEAR VALLEY

Dredging of extensive placers produced 1,168,000 pounds of columbium and tantalum between 1953 and 1959. A total production of $12.5 million in columbium, tantalum, and uranium was realized in those years.

CONCLUSION

During four decades of prospecting and mineral development prior to Idaho's Thunder Mountain gold rush of 1902, miners and community leaders in a variety of isolated camps had learned many different lessons about resource management. They had overcome obstacles common to those faced by farmers, ranchers, and lumbermen, and had dealt with additional problems of their own. Unlike farmers and ranchers, who at least could identify lands superior for their purpose, or lumbermen who had no trouble finding forests with commercial timber stands, miners had two additional problems: locating commercial mineral resources and developing mineral recovery technology, both of which varied greatly in different districts. Except for gold producers, miners also had marketing problems similar to those of other Idaho pioneers. Obstacles to raising investment capital during national financial panics (particularly 1873, 1884, and 1893) plagued lode developers. Declining silver prices in 1888 and 1892 put many of Idaho's miners at a special disadvantage. Otherwise, the miners had investment experience similar to other large-scale transportation, irrigation, and forest resource enterprises in the west. Yet, despite the problems and obstacles, many fast-talking gold and silver promoters still induced unwary capitalists to participate in operations that sounded tremendously promising but that in fact had little or no expectation of success or even of investment salvage. Other unscrupulous resource developers, such as land speculators, had also managed to sell worthless holdings, even though lands useless for farming could at least be identified easily. Most mining properties could not be evaluated satisfactorily.

The exceptional opportunities to gain fabulous wealth accounted for a special kind of excitement in the search for gold and silver, or even for lead and copper. Eager bonanza hunters almost never succeeded in selling high-grade claims for a fortune, and many competent prospectors were too restless to quit prospecting even if they enjoyed a degree of success in selling their claims. Most free-milling gold and silver had been noticed when lead-silver began to attract more interest after 1880. Neither silver nor base metals could be traced from placer deposits very easily, and values of lead-silver or copper ores could not be tested at depth by simple panning methods. So new mineral tests and prospecting procedures had to be learned. Prospectors managed to recognize new kinds of mineralized outcrops so that they could get assays which would identify valuable lodes. Until about 1900, they always had more ridges and formations to explore. By that

time — as indicated by the rush to Idaho's Thunder Mountain, a remote district lacking in mineral wealth — few large gold and silver properties remained undiscovered. Prospectors found enough small, rich orebodies to keep up a mining fever through another decade after 1900. An occasional district turned out to be of major importance, but prior to extensive and costly development, major lodes could not be distinguished from minor producers. Development often was neglected, and rich but minor properties appeared to exceed their actual value. So mining excitement continued, often without the minerals to justify it.

Nineteenth century lode miners in Idaho had another special problem, shared with railroad builders but with few other enterprises. Industrial labor, much of it employed in hazardous or disagreeable locations, demanded economic stability and safer working conditions; and this was achieved through the formation of unions. During times of national economic adversity, labor unrest and strikes disrupted many Idaho camps, such as Wood River in 1884. National railroad strikes, such as those of 1886, also troubled Idaho's mining communities. Idaho's economy had always been subject to the financial and labor trends of the nation, and this dependence became much more evident in later years.

Regardless of economic difficulties characteristic of large-scale western enterprises, Idaho's miners joined enthusiastically in resource development. They appropriated public lands for mineral rights just as farmers, ranchers, and loggers did. Farmers had obtained a federal homestead act in 1862 to provide a legal basis for their taking over what land they needed, and miners got a similar federal concession in 1872. Ranchers and lumbermen, unable to obtain suitable legislation, simply went ahead and used public lands anyway, although they eventually ran into restrictive statutes that farmers and miners had managed to avoid. Miners and stockraisers sometimes failed to avoid friction, but their claims wars and sheep and cattle wars were generally fought out in court litigation with only occasional violence attending their utilization of Idaho's resources.

Late nineteenth century mining in Idaho, as in many other western mineral areas, gained well-deserved notoriety for risk and for reckless financial operation. Many lode properties, which could be developed only with large capital investment and advanced technology, were promoted with excessive sales pressure substituted for sound management. Following well-established English precedents in Cornwall and other important mining areas where swindlers had begun successful promotions by erecting mills without bothering to find mines at all, managers were tempted to neglect or dispense with the development essential to sound operation. Even if they followed appropriate engineering procedures, lode miners still faced considerable practical risk in extracting and processing ore. In many places veins declined in value at depth, and miners too frequently continued to operate long after they should have quit. (The great Comstock lode bonanza, encountered at surprising depth—a discovery not typical of most western mining districts, deceived many investors into seeking similar wealth in other places where that extent of riches was not to be found.) Other properties had ores which could not be processed for many years until technological problems were solved. Price fluctuations for lead, silver, and other metals interfered with efficient operations. Miners as well as investors were affected adversely by such unmanageable risks. Mining operations would start and then shut down erratically, so that miners often were left unemployed at very awkward times. Along with investors they had to absorb their own severe losses resulting from unemployment and from the costs in moving to other camps where similar risks were to be encountered again. As a mining territory, Idaho (like California and Nevada) attracted a frontier population accustomed to taking risks but eager to gain wealth on a scale that ordinary occupations normally failed to provide.

Oregon, by contrast, had been settled by pioneers who wanted as stable a farming economy as could be developed, yet that kind of low risk situation did not always materialize either. Markets often were unavailable for the crops that could be harvested in abundance. Some farming areas had dependable moisture where crop failures rarely occurred. When southern Idaho finally had sufficiently large irrigation projects, the access to water reduced the hazard in agriculture. But aside from Mormon irrigation cooperatives, which were not utilized in Idaho mining, large canal company projects entailed more risk and failure than mining. Later, state and federal government participation (another development and management device not resorted to for mining until after 1940) eventually enabled farmers to overcome construction risks for major reclamation enterprises. Aside from small suppliers of farm products to mining camps (with operating risks often as low as in placer mining, a relatively safe endeavor), nineteenth century farmers trying to irrigate large tracts of land had to deal with risks at least as great as those confronting lode miners. Erratic price and marketing problems confronted farmers and miners alike, providing additional hazards to those fortunate enough to have been productive.

Other major nineteenth century enterprises fared no better. Stockraisers enjoyed an era of dramatic success, but disastrous winters and overgrazing proved ruinous to many large operations after 1888. Commercial logging, only beginning in Idaho during that time, fared no better. Even after improved transportation and marketing conditions supported a large forest products industry, Idaho lumbermen had to overcome substantial risks. These arose from fire hazard, transportation difficulties, and marketing problems. Failures and losses were at least comparable with those of lode mining companies. Relatively speaking, mining risks in Idaho were not out of line for an economy in which hazard abounded.

In spite of a fair share of hazard and loss, Idaho's nineteenth century miners founded substantial communities which supported a farming and ranching commonwealth that qualified for state admission in 1890. Without a mining economy, Idaho's agricultural settlement would have been retarded for many years. Aside from Mormon expansion from Utah, which came independently of mining development, southern Idaho would have offered little attraction to farmers and loggers until rail transportation became available. North Idaho would have remained part of Washington, and nothing resembling Idaho would have been likely to materialize. Along with a number of other western states, Idaho developed as a mining territory with distinctive characteristics resulting from its gold rush origin. To understand modern Idaho, a complex pattern of mining antecedents has to be investigated and explained.

Historic Value of Metal Production for Idaho, 1860-1980
Cumulative Totals by Mining Area

Atlanta	$ 16,000,000	Miller's Camp-Secesh	500,000
Banner	3,000,000	Mineral City	800,000
Bay Horse-Clayton	42,000,000	Muldoon	200,000
Bear Valley	12,800,000	Neal	2,000,000
Big Creek	400,000	Newsome-Golden	2,940,000
Blackbird (Cobalt)	49,000,000	Orogrande	640,000
Boise Basin	60,000,000	Owyhee	90,000,000
Boise Ridge	428,000	Palouse	340,000
Boise River	450,000	Patterson	10,000,000
Buffalo Hump	540,000	Pearl	400,000
Cariboo Mountain	1,200,000	Pend d'Oreille	2,000,000
Clark's Fork	2,500,000	Pierce	8,000,000
Coeur d'Alene	3,845,732,000*	Porthill	4,500,000
Deadwood	1,200,000	Rocky Bar-Pine	6,090,000
Dixie (South Fork Clearwater)	1,500,000	Salmon River Bars	2,500,000
Elk City	16,000,000	Seafoam-Greyhound	400,000
Era and Martin	400,000	Seven Devils-Heath	2,800,000
Florence	9,600,000	South Mountain	1,900,000
Germania-Livingston	650,000	Stanley	400,000
Gibbonsville	2,000,000	Stibnite	53,000,000
Gilmore	11,600,000	Thunder Mountain	500,000
Hailey Gold Belt	1,000,000	Ulysses	600,000
Leadore	300,000	Vienna-Sawtooth City	800,000
Leesburg	5,420,000	Viola	2,500,000
Lemhi	1,940,000	Warren	16,120,000
Little Lost River	2,000,000	Weiser Mercury	8,000,000
Little Smoky	1,200,000	Wood River	62,000,000
Long Valley	3,500,000	Yankee Fork	12,000,000
Loon Creek	1,200,000	Yellow Jacket	400,000
Mackay-Copper Basin	15,000,000		
Marshall Lake	2,000,000	TOTAL	4,402,890,000

*Value of metal production in the Coeur d'Alene mining district rose to $4.2 billion by the end of 1982.

Many of these totals are based largely upon reliable sources (usually Bureau of Mines or other governmental compilations), but some are of unknown accuracy. Most lode and dredge production figures are reliable, and almost all Idaho metal production is of those kinds. Less than 2% of the total production of $4,402,890,000 are from sources of uncertain accuracy. This table must be used with great caution, since mineral prices — even for gold — varied greatly over the century the figures cover. Inflation has weighted the period since 1940 very heavily. Gold and silver prices after 1976 have fluctuated ten or twenty times as much. Boise Basin's gold values, for example, recently have risen to more than one or two billion dollars. Other major gold districts (Elk City, Florence, Leesburg, Pierce, Rocky Bar, Warren and Yankee Fork) have increased similarly. Silver prices have varied much more. Assignment of smelting values and federal support prices also affects some of these totals (such as Blackbird) to a marked degree.

Index

www.ingramcontent.com/pod-product-compliance
Lightning Source LLC
Chambersburg PA
CBHW062043090426
42740CB00016B/3001